PHYSICS THROUGH THE 1990s

Elementary-Particle Physics

Elementary-Particle Physics Panel

Physics Survey Committee

Board on Physics and Astronomy

Commission on Physical Sciences,
Mathematics, and Resources

National Research Council

NATIONAL ACADEMY PRESS
Washington, D.C.   1986

NATIONAL ACADEMY PRESS 2101 Constitution Avenue, NW Washington, DC 20418

NOTICE: The project that is the subject of this report was approved by the Governing Board of the National Research Council, whose members are drawn from the councils of the National Academy of Sciences, the National Academy of Engineering, and the Institute of Medicine. The members of the committee responsible for the report were chosen for their special competences and with regard for appropriate balance.

This report has been reviewed by a group other than the authors according to procedures approved by a Report Review Committee consisting of members of the National Academy of Sciences, the National Academy of Engineering, and the Institute of Medicine.

The National Research Council was established by the National Academy of Sciences in 1916 to associate the broad community of science and technology with the Academy's purposes of furthering knowledge and of advising the federal government. The Council operates in accordance with general policies determined by the Academy under the authority of its congressional charter of 1863, which establishes the Academy as a private, nonprofit, self- governing membership corporation. The Council has become the principal operating agency of both the National Academy of Sciences and the National Academy of Engineering in the conduct of their services to the government, the public, and the scientific and engineering communities. It is administered jointly by both Academies and the Institute of Medicine. The National Academy of Engineering and the Institute of Medicine were established in 1964 and 1970, respectively, under the charter of the National Academy of Sciences.

The Board on Physics and Astronomy is pleased to acknowledge generous support for the Physics Survey from the Department of Energy, the National Science Foundation, the Department of Defense, the National Aeronautics and Space Administration, the Department of Commerce, the American Physical Society, Coherent (Laser Products Division), General Electric Company, General Motors Foundation, and International Business Machines Corporation.

Library of Congress Cataloging in Publication Data

National Research Council (U.S.). Elementary-
    Particle Physics Panel.
    Elementary-particle physics.

    (Physics through the 1990s)
    Bibliography: p.
    Includes index.
    1. Particles (Nuclear physics)—Congresses.
2. National Research Council (U.S.). Elementary-
Particle Physics Panel—Congresses. I. Title.
II. Series.
QC793.N36   1985   539.7'2   85-15210
ISBN 0-309-03576-7

Printed in the United States of America

PANEL ON ELEMENTARY-PARTICLE PHYSICS

MARTIN L. PERL, Stanford Linear Accelerator Center, *Chairman*
CHARLES BALTAY, Columbia University
MARTIN BREIDENBACH, Stanford Linear Accelerator Center
GERALD FEINBERG, Columbia University
HOWARD A. GORDON, Brookhaven National Laboratory
LAWRENCE W. JONES, University of Michigan
BOYCE D. MCDANIEL, Cornell University
FRANK S. MERRITT, The University of Chicago
ROBERT B. PALMER, Brookhaven National Laboratory
JAMES M. PATERSON, Stanford Linear Accelerator Center
JOHN PEOPLES, JR., Fermi National Accelerator Laboratory
CHRIS QUIGG, Fermi National Accelerator Laboratory
DAVID M. RITSON, Stanford University
DAVID N. SCHRAMM, The University of Chicago
A. J. STEWART SMITH, Princeton University
MARK W. STROVINK, University of California, Berkeley

DONALD C. SHAPERO, National Research Council, *Executive Secretary*

Liaison Members

D. BERLEY, National Science Foundation, Liaison for National
  Science Foundation
L. S. BROWN, University of Washington, Liaison for Division of
  Particles and Fields, American Physical Society
W. A. WALLENMEYER, Department of Energy, Liaison for
  Department of Energy

Panel Editor

W. T. KIRK, Stanford Linear Accelerator Center

iii

iv

vi

Preface

This report on elementary-particle physics is part of an overall survey of physics carried out for the National Academy of Sciences by the National Research Council. The panel that wrote this report had three goals. The first goal was to explain the nature of elementary-particle physics and to describe how research is carried out in this field. The second goal was to summarize our present knowledge of the elementary particles and the fundamental forces. The third goal was to consider the future course of elementary-particle physics research and to propose a program for this research in the United States.

It is the hope and intention of the panel that this volume will be read and found useful outside the physics community. Therefore the text does not assume that the reader has any special knowledge of mathematics or of physics beyond an acquaintance with general notions such as mass and energy. Nor do we assume that the reader has any previous knowledge of the techniques of elementary-particle physics research, namely, accelerators and particle detectors. Indeed we have presented basic introductions to these techniques.

In the last two decades there has been a revolution in our knowledge of elementary-particle physics. We have identified three types of elementary particles—the quarks, the leptons, and the force-carrying particles; we have learned a great deal more about three of the fundamental forces; and the weak force and electromagnetic force have been unified in a beautiful and powerful theory. Major innovations have been made in the technologies of accelerators and of particle detectors. In order to present all of this in a few chapters, we have had to limit ourselves to describing the main ideas and the major experimental and theoretical accomplishments. We apologize to our colleagues for leaving out descriptions or even mention of so much other important and beautiful work in elementary-particle physics.

Elementary-particle physics is an international science, and in describing its content and its methods we have used the work of all the world's elementary-particle physicists. In looking to the future needs and future opportunities of elementary-particle physics we have mostly limited our work and our presentation to the United States. We have done so because this was the charge to the Physics Survey Committee from the National Academy of Sciences of the United States and because we are constituted primarily as a panel of physicists from the United States who are not qualified to speak for physics abroad. Since one of the audiences for this report consists of members of the federal government of the United States who are concerned with science policy, in describing needs and opportunities we have naturally tended to use examples from the elementary-particle physics community in the United States. We hope that our colleagues abroad will understand that this was one of the purposes of the report and will not feel slighted by our inability in this limited space to present more examples from work of the elementary-particle physics community abroad.

The Elementary-Particle Physics Panel acknowledges the help it has had from many physicists who have graciously given their time for discussions on the contents of this volume, who have read and reviewed individual sections, and who have been kind enough to review and make suggestions for the entire volume. We are very grateful to John Ellis of CERN, who attended the early meetings of the Panel and wrote some of the first drafts of this report. We have tried to represent the views of the elementary-particle physics community as a whole, but of course it is only panel members who bear the responsibility for the material in this volume. We thank the Chairman of the Physics Survey Committee, William F. Brinkman, for his guidance, leadership, and wisdom. We express our gratitude to the Staff Director of the Board on Physics and Astronomy, Donald C. Shapero, who was so patient and generous in passing on to us his knowledge and experience of how to represent the views of a scientific community and of how to prepare a report of this nature. Finally, we thank the technical typists and illustrators who so patiently worked and reworked the many drafts of this report: Lydia Beers, Edythe Christianson, and the members of the Publications Office of the Stanford Linear Accelerator Center.

Contents

APPENDIXES

Executive Summary

Elementary-particle physics, the science of the ultimate constituents of matter and the interactions among them, has undergone a remarkable development during the past two decades. A host of new experimental results made accessible by a new generation of particle accelerators and the accompanying rapid convergence of theoretical ideas has brought to the subject a new coherence and has raised new possibilities and set new goals for understanding nature. The progress in particle physics has been more dramatic and more thoroughgoing than could have been imagined at the time of the 1972 survey of physics, *Physics in Perspective* (National Academy of Sciences, Washington, D.C., 1972). Many of the important issues identified in that report have been addressed, and many of the opportunities foreseen there have been realized. As a result, we are led to pose new and more fundamental questions and to conceive new instruments that will enable us to explore these questions.

Elementary-particle physics is the study of the basic nature of matter, energy, space, and time. Elementary-particle physicists seek the fundamental constituents of matter and the forces that govern their behavior. In common with all physicists, they seek the unifying principles and physical laws that determine the material world around us.

The atom, the atomic nucleus, and the elementary particles of which they are composed are too small to be seen or studied directly. Throughout this century, physicists have devised ever more sophisti-

1

cated detection devices to observe the traces of these particles and their constituents. At the same time, they have developed increasingly energetic beams of particles to probe deeply into the structure of matter. Early examples are x rays to probe the electronic structure of the atom and radioactive sources to study the atomic nucleus. Some of the constituents of ordinary matter, notably electrons and protons, are quite stable and easily manipulated in electric and magnetic fields. They can therefore be accelerated to high energies and used as probes to reach the very small distance scale of the fundamental constituents. The colliding of high-energy particles and the analysis of collision products is at the heart of experimental particle physics. For this reason the field is often called *high-energy physics*.

THE REVOLUTION IN PARTICLE PHYSICS

Thirty years ago, ordinary matter was thought to consist of protons, neutrons, and electrons. Experiments were under way to probe the structure of these particles and to study the forces that bind them together into nuclei and atoms. In the course of these experiments, physicists discovered more than a hundred new particles, called *hadrons*, which had many similarities to the proton and the neutron. None of these particles seemed more elementary than any other, and there was little understanding of the mechanisms by which they interacted with one another.

Since that time, a radically new and simple picture has emerged as a result of many crucial discoveries and theoretical insights. It is now clear that the proton, the neutron, and other hadrons are not elementary. Instead, they are composite systems made up of much smaller particles called *quarks*, much as an atom is a composite system made up of electrons and a nucleus. Five kinds of quarks have been established, and initial experimental evidence for a sixth species has been reported.

Unlike the neutron and the proton, the electron has survived the revolution intact as an elementary constituent of matter, structureless and indivisible. However, we now know that there are six kinds of electronlike particles called *leptons*. According to our present understanding, then, ordinary matter is composed of quarks and leptons.

An important difference between quarks and leptons is that a formidable interaction, known as the *strong force*, binds quarks together into hadrons but does not influence leptons. Both quarks and leptons are acted upon by the three other fundamental forces: the electromagnetic force, the weak force responsible for certain radioactive decays, and the gravitational force.

Over the past two decades, great progress has been made in understanding the nature of the strong, weak, and electromagnetic forces. A unified theory of the weak and electromagnetic forces now exists. Its predictions have been dramatically verified by many experiments, culminating in the discovery of the W and Z particles in 1983. These carriers of the weak force are analogous to the *photon*, the carrier of the electromagnetic interaction, whose existence was established in the 1920s. In addition, there is indirect but persuasive evidence for particles called *gluons*, the carriers of the strong force. Strong, weak, and electromagnetic interactions all are described by similar mathematical theories called *gauge theories*. At this time, the role played by the gravitational force in elementary-particle physics is unclear. We have not been able to measure directly any effect of gravity on the collisions of elementary particles.

With the identification of quarks and leptons as elementary particles, and the emergence of gauge theories as descriptions of the fundamental interactions, we possess today a coherent point of view and a single language appropriate for the description of all subnuclear phenomena. This development has made particle physics a much more unified subject, and it has also helped us to perceive common interests with other specialties. One important by-product of recent developments in elementary-particle physics has been a recognition of the close connection between this field and the study of the early evolution of the universe from its beginning in a tremendously energetic, primordial explosion called the big bang. Particle physics provides important insights into the processes and conditions that prevailed in the early universe, and deductions from the current state of the universe can in turn give us information about particle processes at energies that are too high to be produced in the laboratory, energies that existed only in the first instants after the primordial explosion.

WHAT WE WANT TO KNOW

Developments in elementary-particle physics during the past decade have brought us to a new level of understanding of physical laws. This new level of understanding is often called the *standard model* of elementary-particle physics. As usual, the attainment of a new level of understanding refocuses attention on old problems that have refused to go away and raises new questions that could not have been asked before. The quark model of hadrons and the gauge theories of the strong, weak, and electromagnetic interactions organize our present knowledge and provide a setting for going beyond what we now know.

Although the standard model provides a framework for describing elementary particles and their interactions, it is incomplete and inadequate in many respects. We still do not understand what determines the basic properties of quarks and leptons, such as their masses. Nor do we understand fully how the differences between the massless electromagnetic force carrier, the photon, and the massive carriers of the weak force, the *W* and *Z* particles, arise. Existing methods for dealing with these questions involve the introduction of many unexplained numerical constants into the theory—a situation that many physicists find arbitrary and thus unsatisfying. Physicists are actively seeking more complete and fundamental answers to these questions.

Another set of questions goes beyond the existing synthesis. For example, how many kinds of quark and lepton are there? How are the quarks and leptons related, if they are related? How can the strong force be unified with the already unified electromagnetic and weak forces?

Then there are questions related to our overview of elementary-particle physics. Are the quarks and leptons really elementary? Are there yet other types of forces and elementary particles? Can gravitation be treated quantum mechanically, as are the other forces, and can it be unified with them? More generally, will quantum mechanics continue to apply as we probe smaller and smaller distances? Do we understand the basic nature of space and time?

THE TOOLS OF ELEMENTARY-PARTICLE PHYSICS

Elementary-particle physics progresses through a complicated interaction between experiment and theory. As experimental work produces new data, theory is tested by the data, and theory is used to organize the data. Sometimes theoretical insight leads to new experiments; sometimes an experiment produces surprising new data that upset currently accepted theories. Patient accumulation of data may lead to paradoxes that cannot be resolved without major revision of theoretical ideas. And sometimes experimenters may seek new entities, such as free quarks or magnetic monopoles, which do not fit known patterns. In the end, physics is an experimental science, and it is only experiment and observation that can tell us if we are right or wrong.

Most experiments in our field are carried out by the use of accelerators, which produce beams of high-energy particles. These beam particles collide either with a stationary target (a "fixed-target" experiment) or with another beam of particles. Accelerators in which two beams of particles collide are called colliders. Either in fixed-target experiments or in colliders, the results of the collisions are recorded with devices, often complex, called particle detectors. Accelerators and particle detectors are the main tools of elementary-particle physics. Through the years invention, research, and development have led to major innovations and vast improvement in the technology of accelerators and detectors. In turn, these tools are fundamental to experimental progress in our field.

The fixed-target experiments of the past two decades have contributed much to our knowledge. Examples of these experimental results are the demonstration that neutrons and protons are composed of quarks, one of the two simultaneous discoveries of the fourth (or charmed) quark, the discovery of the fifth (or bottom) quark, and the discovery of the violation of what were thought to be fundamental symmetries in time and space. Fixed-target experiments have accumulated a large body of data that has led to the systematic understanding of the interactions of hadrons.

Experiments utilizing colliders have become increasingly prominent because more of the beam energy is available to the fundamental collision processes. The extension of colliding-beam accelerator technology was led by the development of electron-positron and proton-proton colliders and by other basic advances in that technology. Experiments at electron-positron colliders have given us the shared discovery of the charmed quark; the discovery of the unexpected new "relative" of the electron—the tau lepton; the discovery of intense jets of hadrons; and much of the evidence for the theory that the strong force is mediated by the gluon particle. Recently the development of the proton-antiproton collider contributed substantially to particle physics by making possible the discovery of the carriers of the weak force—the W and Z particles. This development confirms an expanded future role for proton-proton and proton-antiproton colliders.

Most of the discoveries described above were made possible only through the building of new high-energy particle accelerators. This is most evident in the discoveries of the new massive particles, such as the W, the Z, the heavy quarks, and the new lepton. Higher-energy accelerators in the future will similarly open up the possibility of discovering new fundamental particles of still higher mass.

Progress in elementary-particle physics also depends on studying rare or unusual collisions. Therefore it is important to have very intense beams of particles to produce the rare events within a background of less-interesting phenomena. Thus, both intensity and energy are critical parameters of high-energy accelerators.

THE FUTURE OF ELEMENTARY-PARTICLE PHYSICS IN THE UNITED STATES

Elementary-particle physics is perhaps the most basic of the sciences; it interacts with many other areas of physics and astronomy; it develops, stimulates, and uses new technologies. Two decades ago the United States was the dominant force in elementary-particle research. Gradually other regions, particularly Western Europe and Japan, have increased their elementary-particle physics programs until together they equal or exceed the U.S. program in personnel, financial support, and scientific accomplishment. This is as it should be, since science is a worldwide endeavor. International participation leads to innovation in accelerator and detector technology, to an interchange of ideas, and to a more rapid pace of discovery. Indeed, many of the most important recent discoveries have been made in Europe. This report includes recommendations for the future U.S. program in this field that are intended to exploit the scientific opportunities before us and to permit us to maintain a competitive role in the forefront of this science.

The program for the future of the field embodied in our recommendations has emerged from an intense discussion within the community of elementary-particle physicists. During the past 3 years physics study groups and federal advisory panels have considered several different initiatives for new facilities. They have also considered the balance between support of existing facilities and construction of new facilities. Ultimately the choice was determined by the belief that new phenomena that are crucial to the understanding of fundamental problems will be discovered in the tera-electron-volt (TeV) mass range. This region cannot be reached either by existing accelerators or by the accelerators now under construction. The successful conclusion of the long and difficult development of superconducting magnet technology makes a large new machine a feasible and timely choice. Our recommendations form a plan that has as its keystone the construction of a very-high-energy superconducting proton-proton collider, the Superconducting Super Collider (SSC).

RECOMMENDATIONS FOR UNIVERSITY-BASED RESEARCH GROUPS AND USE OF EXISTING FACILITIES IN THE UNITED STATES

The community of elementary-particle physicists in the United States consists of about 2400 scientists, including graduate students, based in nearly 100 universities and 6 national laboratories. They work together in groups frequently involving several institutions. It is their experiments, their calculations, their theories, their creativity that are at the heart of this field. The diversity in size, in scientific interests, and in styles of experimentation of these research groups are essential to maintaining the creativity in the field. *Therefore we recommend that the strength and diversity of these groups be preserved.*

Most elementary-particle physics experiments in the United States are carried out at four accelerator laboratories. Two fixed-target proton accelerators are now operating: the 30-GeV Alternating Gradient Synchrotron at the Brookhaven National Laboratory and the 1000-GeV superconducting accelerator, the Tevatron, at the Fermi National Accelerator Laboratory. Cornell University operates the electron-positron collider CESR. The Stanford Linear Accelerator Center operates a 33-GeV fixed-target electron accelerator, which also serves as the injector for two electron-positron colliders, SPEAR and PEP. In addition, some elementary-particle physics experiments are carried out at medium-energy accelerators that are devoted primarily to nuclear physics.

Experimentation at the four accelerator laboratories requires complex detectors that are often major facilities in their own right. The equipment funds for major detectors and the operating funds for the accelerators have been insufficient to permit optimum use. Because accelerator laboratories necessarily have large fixed costs, the productivity of the existing accelerator facilities can be increased considerably by a modest increase in equipment and operating funds. *We recommend fuller support of existing facilities.*

RECOMMENDATIONS FOR NEW ACCELERATOR FACILITIES IN THE UNITED STATES

The capability of two existing accelerators in the United States is now being extended by adding collider facilities to each of them. A 100-GeV electron-positron collider, which uses a new linear collider principle, is now being constructed at the Stanford Linear Accelerator Center. The Tevatron at the Fermi National Accelerator Laboratory is being completed so that the superconducting ring can also be operated as a 2-TeV proton-antiproton collider. *We recommend continued support for the completion of these new colliders on their present schedule. In addition, we recommend that their experimental facilities and programs be fully developed.*

The U.S. elementary-particle physics community is now carrying out an intensive research, development, and design program intended to lead to a proposal for the very-high-energy, superconducting proton-proton collider, the SSC. This new collider will be based on the accelerator principles and technology that have been developed at several national laboratories and in particular on the extensive experience with superconducting magnet systems that has been gained at the Fermi National Accelerator Laboratory and Brookhaven National Laboratory. The SSC energy would be about 20 times greater than that of the Tevatron collider. This higher energy is needed in the search for heavier particles, to find clues to the question of what generates mass, and to test new theoretical ideas. Our current ideas predict a rich world of new phenomena in the energy region that can be explored for the first time by this accelerator. Furthermore, history has shown that the unexpected discoveries made in a new energy regime often prove to be the most exciting and fundamentally important for the future of the field. On its completion this machine will give the United States a leading role in elementary-particle physics research. *Since the SSC is central to the future of elementary-particle physics research in the United States, we strongly recommend its expeditious construction.*

RECOMMENDATIONS FOR ACCELERATOR RESEARCH AND DEVELOPMENT

Since accelerators are the heart of most elementary-particle experimentation, physicists are continuing research and development work on new types of accelerators. Indeed, technological innovation in accelerators has been the driving force in extending the reach of high-energy physics. An important part of this work is concerned with extending the electron-positron linear collider to yet higher energies. One of the purposes of the construction of the Stanford Linear Collider is to serve as a demonstration and first use of such a technology. Advanced accelerator research is also exploring new concepts, based on a variety of technologies, that may provide the basis for even more powerful accelerators, perhaps to be built in the next century. Such research also leads to advances in technology for accelerators used in industry, medicine, and other areas of science such as studies based on synchrotron radiation. *We recommend strong support for research and development work in accelerator physics and technology.*

RECOMMENDATIONS FOR THEORETICAL RESEARCH IN PARTICLE PHYSICS

Theoretical work in elementary-particle physics has provided the intellectual foundations that motivate and interconnect much experimental research. Elementary-particle theorists have also played an important role in forging links with other disciplines, including statistical mechanics, condensed-matter physics, and cosmology. Theoretical physicists make vital contributions to university research programs and to the education of students who will enter all branches of physics.

We recommend that the existing strong support for a broad program of theoretical research in the universities, institutes, and national laboratories be continued. A new element of theoretical research is the increasing utilization of computer resources, which has spurred the development and implementation of new computer architectures. This trend will require the evolution of new equipment-funding patterns for theory.

RECOMMENDATIONS FOR NONACCELERATOR PHYSICS EXPERIMENTS

It is appropriate that some fraction of the particle-physics national program be devoted to experiments and facilities that do not use accelerators. These experiments include the searches for proton decay by using large underground detectors, the use of cosmic rays to explore very-high-energy particle interactions, the measurements of the rate of neutrino production by the Sun, and the use of nuclear reactors to study subtle properties of neutrons and neutrinos. There are also diverse experiments that search for evidence of free quarks, magnetic monopoles, and finite neutrino mass. Still other classes of experiments overlap the domain of atomic physics; these include exquisitely precise tests of the quantum theory of electromagnetism, studies of the mixing of the weak and electromagnetic forces in atomic systems, and searches for small violations of fundamental symmetry principles through a variety of different techniques. Many of these are small-scale laboratory experiments. Some provide a means of probing an energy scale inaccessible to present-day accelerators.

The value of these experiments is substantial. *They will continue to play a vital role that is complementary to accelerator-based research, and we recommend their continued support.*

RECOMMENDATIONS FOR INTERNATIONAL COOPERATION IN ELEMENTARY-PARTICLE PHYSICS

Our program should be designed to preserve the vigor and creativity of elementary-particle physics in the United States and to maintain and extend international cooperation in the discipline. *We recommend four guidelines for such a balanced program.* First, the continued vitality of American elementary-particle physics requires that there be pre-eminent accelerator facilities in the United States. The use of accelerators developed by other nations provides a needed diversity of experimental opportunities, but it does not stimulate our nation's technological base as do the conception, construction, and utilization of innovative facilities at home. The SSC will be a frontier scientific facility, and the technological advances stimulated and pioneered by its design and construction will serve the more general societal goals as well. Second, the most productive form of cooperation with respect to accelerators is to develop and build complementary facilities that allow particle physics to be studied from different experimental directions. Third, the established forms of international cooperation, including the use of accelerators of one nation by physicists from another nation, should be continued. Fourth, looking beyond the program proposed in this report, there should be further expansion of international collaboration in the planning and building of accelerator facilities.

CONCLUSION

We believe that the implementation of the recommendations made above will enable the United States to maintain a competitive position in the forefront of elementary-particle physics research into the next century. Central to this future is the construction of the SSC, the very-high-energy proton-proton collider using superconducting magnets.

1

Introduction

Over the last two decades our understanding of the fundamental nature of matter has undergone a revolution. The ideas we learned as students have been superceded by concepts at once simpler and more elegant. The more than 100 different kinds of particles identified in the early 1960s are now known to be made up of only a few different kinds of more elementary (simpler) particles called quarks. Seemingly unrelated particles such as quarks and electrons are found to be related. And two apparently different forces have been shown experimentally and theoretically to be simply different manifestations of the same more fundamental electroweak force.

This revolution in elementary-particle physics is the result of the continuing interplay between new theories and new experimental results. Almost all of these experiments have used accelerators, the machines that produce the high-energy particles that are needed to study matter on the smallest scale. These machines have included both traditional accelerators, those in which a beam of high-energy particles strikes a stationary target, and also the newer colliding-beam machines, accelerators in which two beams of high-energy particles collide head on.

We are now in a position to look for the answers to yet more basic questions: What determines the properties of the elementary particles? How many different kinds are there? How are they related to each other? Are the other forces in nature also simply different aspects of a single, truly fundamental force? The new colliders can provide the immense energies needed to answer some of these questions, while others can be attacked by more traditional methods. Thus the next several decades offer us the opportunity to continue the remarkable progress of the recent past. This report is about that progress and also about the opportunities for future progress.

ELEMENTARY-PARTICLE PHYSICS

Elementary-particle physics is the study of the basic nature of matter, of force, of energy, of time, and of space. We seek to discover the simplest constituents of matter, which we call the elementary particles, and we seek to understand the basic forces that operate between them. Above all, we seek the unifying principles and physical laws that will give us a rational and predictive picture of the elementary particles and the basic forces that constitute our world.

Elementary particles are very small, much smaller than atoms; hence this is the physics of the very small. The size of the objects studied in this field compared with those studied in other areas of physics is sketched in Figure 1.1. Elementary particles are too minute to see or study directly. We examine them and make new types of particles by colliding particles together at high energies. High energies are needed because the elementary particles are very small and very hard; it takes a great deal of energy to penetrate them or to break them up. Colliding two particles together at high energy and studying the results of that collision is the heart of elementary-particle physics experiments. Thus this field of research is also called high-energy physics.

In the last decade, close connections have been established between particle physics and astrophysics. Hence elementary-particle physics is also connected with very-large-scale phenomena.

WHAT WE KNOW

During the past two decades, particle-physics research has cleared away much of the underbrush that had concealed from us the world of elementary particles. As shown in Figure 1.2, we now know that there are three basic families of elementary particles: the quarks, the leptons, and the force-carrying particles. Some of these particles can only be produced in the laboratory or occur very rarely in nature. But others make up the matter of our everyday world. Thus the atoms that make up all matter consist of electrons moving in orbits around the atomic nucleus. The electron is one of the leptons, and the nucleus consists of protons and neutrons that in turn are made up of quarks.

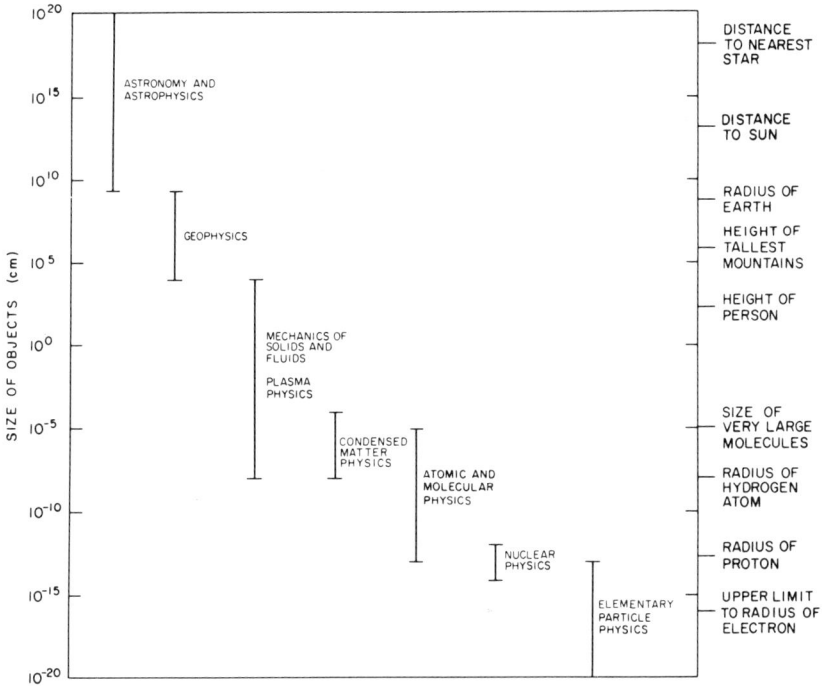

FIGURE 1.1 The different subfields of physics study parts of nature that are very different in size. Elementary-particle physics studies the smallest objects in nature, objects that are smaller than 10^{-13} centimeter.

Figure 1.2 also shows the four known basic forces. Two of these have been known for hundreds of years: the forces of electromagnetism and of gravitation. The other two forces were discovered in the twentieth century. One is the strong or nuclear force that holds the atomic nucleus together, and the other is the weak force that operates in many forms of radioactivity. One of the goals of the physicist is to find out if these four forces can be derived from an even more basic, single, unified force. Significant progress has been made in this direction in the last two decades, as we now know that the electromagnetic and weak forces are two manifestations of a single underlying force.

Thus our present knowledge of elementary-particle physics can now be organized in a simple and elegant way. Chapters 2 and 3 describe this present knowledge in some detail.

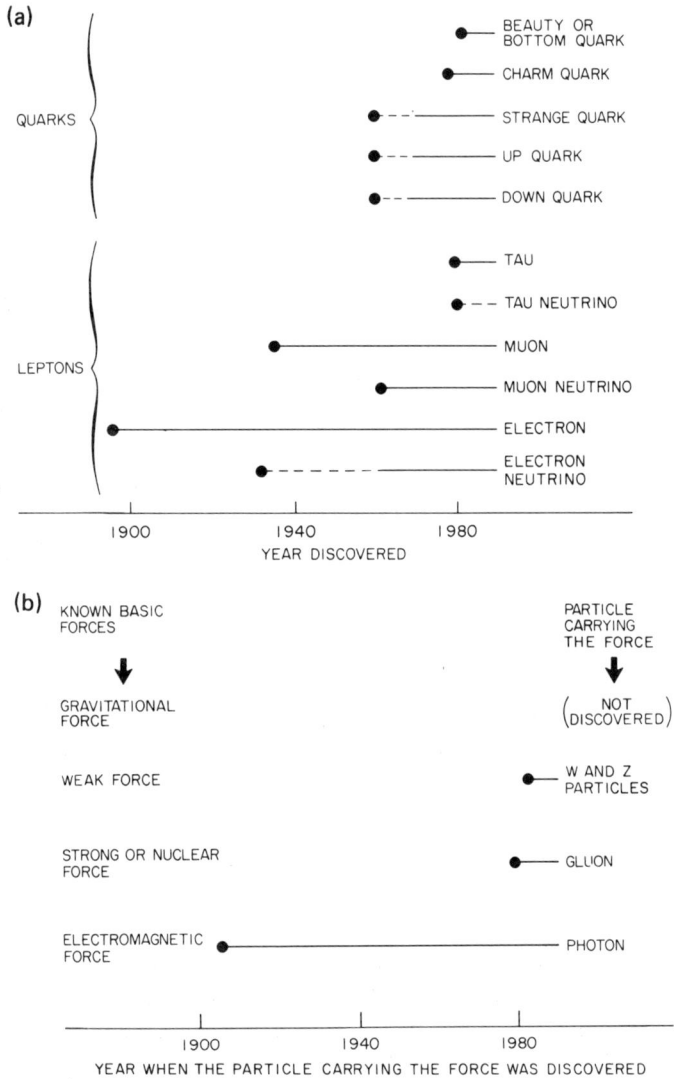

FIGURE 1.2   (a) History of the discovery of the leptons and quarks. The dashed line means that there was strong indirect evidence for the existence of the particle but that the particle itself had not been directly identified. For example, there was strong indirect evidence for the existence of neutrinos before 1940, but the electron neutrino was not identified directly until the 1950s. After this report was completed, initial evidence was reported for the existence of the sixth or top quark. (b) Each of four basic forces is believed to be carried by different elementary particles. This figure identifies the particles that carry the weak, strong, and electromagnetic forces and shows when these particles were discovered. The gravitational force should also be carried by a particle, called the graviton, but at present there is no indirect or direct evidence for its existence.

WHAT WE WANT TO KNOW

Now that we have discovered three basic types of elementary particles, we are finally able to attack some of the central questions that have for so long intrigued and baffled the physicist. As an example, what determines the most basic property of a particle—its mass? The masses of the different kinds of particles vary enormously: the heaviest known quark has more than ten times the mass of the lighter quarks, while the tau lepton has about 3500 times the mass of the electron.

Another important question concerns whether there are more kinds of elementary particles waiting to be discovered. Or, alternatively, can we find a unifying principle that connects the known particles and tells us that there are no more? As already mentioned, we also seek a unifying principle for the basic forces. In Chapter 4 the questions that intrigue and even haunt us are discussed.

THE TOOLS OF ELEMENTARY-PARTICLE PHYSICS

Elementary-particle physics progresses through a complicated interaction between experiment and theory. As experimental work produces new data, theory is tested by the data, and theory is used to organize the data. Sometimes a flash of theoretical insight leads to new experiments; at other times an experiment unexpectedly produces surprising new data and upsets currently accepted theories. Thus experiment and theory are two kinds of tools of elementary-particle physics. In Chapters 5 and 6, we emphasize the experimental tools. We do this because the size and complexity of these tools, particularly of the accelerators, is a special quality of this field. But more fundamentally, physics is an experimental science, and in the end it is only experiments that can tell us if our ideas are right or wrong. Almost all experiments in this field are carried out by using an accelerator to produce high-energy particles, allowing those particles to collide, and then using an apparatus called a detector to find out what has come out of the collision. In the traditional arrangement, a beam of high-energy particles produced by an accelerator strikes a stationary target. Much of the experimental progress in the last two decades has come from such fixed-target experiments. Examples are the demonstration that protons and neutrons are made of quarks, one of the discoveries of the *c* quark, the discovery of the *b* quark, and the discovery of the still mysterious CP violation effect.

Increasingly, however, particle colliders have come to play a dominant role in contributing new knowledge to the field. In such accelerators, two beams of particles collide head on, and this produces much higher energy in the collision than is available in fixed-target accelerators of the same size. Colliding-beam accelerator technology has been paced by the development of electron-positron colliders and proton-proton colliders. Experiments at electron-positron colliders have given us the co-discovery of the *c* quark, the discovery of charmed particles, the completely unexpected discovery of the tau lepton, the discovery of jet structure in particle production, and much of the evidence for the idea that the strong or nuclear force is carried by the gluon particle. Experiments at a proton-proton collider studied the details of the interactions of quarks and gluons. Recently a proton-antiproton collider has begun to contribute substantially to particle physics. The most notable contribution is the discovery in 1983 of the *W* and *Z* particles that carry the weak force.

Our understanding of the physics of accelerators, together with inventions in accelerator technology, has now reached the point that we can substantially increase the energy reached by colliding-beam accelerators. Higher-energy electron-positron colliders and antiproton-proton colliders are now under construction in the United States, Europe, Japan, and the Soviet Union. The antiproton-proton collider being constructed in the United States will be the highest-energy collider in the world when completed in about 1985. The electron-positron collider under construction in the United States uses a new collider technology, called a linear collider, rather than the conventional circular collider. Construction has begun in Europe on an electron-proton collider, something that has never been done before. More information on accelerators is given in Chapter 5.

In this Introduction we have not discussed the kinds of elementary-particle research that do not use high-energy physics accelerators. While not by any means the majority of experiments, such experiments are important in our field. Some use lower-energy accelerators intended for nuclear-physics studies, and some use cosmic rays. Some are conducted in an ordinary laboratory setting, while others are carried out on mountain tops or deep underground. The highlights of this work are given in Chapters 6 and 7.

THE FUTURE TOOLS OF ELEMENTARY-PARTICLE PHYSICS

Experimental investigation of some of the fundamental questions in elementary-particle physics requires energies higher than those provided by any accelerators now in operation or under construction anywhere in the world. For this reason the U.S. elementary-particle physics community is now preparing a proposal for a very-high-energy, superconducting proton-proton collider, the Superconducting Super Collider (SSC). It would be based on the accelerator principles and technology that were developed in connection with the construction of the Tevatron and on other extensive work on superconducting magnets in the United States. This proposed collider would have an effective energy range about 60 times higher than that of any collider now in operation. Not only is the SSC needed to answer some of the questions that we face in elementary-particle physics, but in addition such a large increase in energy will open up new regions of elementary-particle physics to be explored.

Since accelerators are at the heart of elementary-particle experimentation, there is extensive research and development work on new types of accelerators and higher-energy accelerators. An important part of this work concerns extending the electron-positron linear collider to yet higher energies. It seems quite likely that technology can be developed to build a very-high-energy electron-positron collider. Since the physics that can be done at such a collider is mostly complementary to that which can be done at a proton-proton collider, the elementary-particle physicist would hope to see both types of collider in operation eventually.

2

What Is Elementary-Particle Physics?

Elementary-particle physics deals with questions first recorded by the philosophers of classical Greece. What is the basic nature of the material world around us? What are the simplest, the most elementary, kinds of matter? What are the basic forces that operate in our material world?

Although these are very old questions, it was not until about four centuries ago that scientists began to make progress in trying to answer them. Some of the first answers came with the discovery of certain of the basic forces in nature: the gravitational force, the electrical force, and the magnetic force. It was not until the middle of the nineteenth century that it was discovered that the electric and magnetic forces are in fact two different aspects of the force that we now call electromagnetism.

Progress in the study of the basic nature of matter itself also came slowly. Indeed, it was not until the last decade of the nineteenth century that the first of the particles that we now call elementary was discovered; this was the electron. In the next six decades only a few more kinds of truly elementary particles were discovered: the muon, the neutrinos, and the photon. It is just in the last two decades that tremendous progress has been made in our field—that we have been able to understand the families of elementary particles and have been able to get for the first time a full view of the basic nature of matter.

This chapter is devoted to introducing the fundamental ideas of particle physics that have been developed over the last 50 years. We will attempt to present these ideas in a way that does not require a previous knowledge of high-energy physics nor of mathematics. Chapter 3 will explore our present picture in somewhat greater detail and will describe in particular how these ideas have been developed and verified over the last two decades.

WHAT IS AN ELEMENTARY PARTICLE?

We call a piece of matter an elementary particle when it has no other kinds of particles inside of it and no subparts that can be identified. We think of an elementary particle as occupying no room in space; indeed, we often think of it as a point particle.

How do we know whether a particle is elementary? We know only by experimenting with it to see if it can be broken up or by studying it to determine if it has an internal structure or parts. This is illustrated in Figure 2.1. We know that molecules are not elementary because they can be broken up into atoms by chemical reactions or by heating or by other means. Nor are atoms elementary: they can be broken up into electrons and nuclei by bombarding the atom with other atoms or with light rays. Nor is the nucleus elementary: by bombarding nuclei with high-energy particles or with high-energy light rays called gamma rays, the nucleus can also be broken up into protons and neutrons.

For about 50 years physicists considered the neutron and proton to be elementary, but in the last two decades we have found that these particles themselves are made up of yet simpler particles called quarks. That is, protons and neutrons have other particles inside of them, hence they are not elementary. However, we have no evidence as yet that the neutron and proton can actually be broken up into these individual quarks; this is a subtle point and is discussed later.

What about the electron, the other constituent part of the atom? Despite all of our experiments and all of our probing of the electron, we have not succeeded in breaking up an electron, and we cannot find any evidence that electrons have internal parts or structure. This is why we call the electron an elementary particle.

How Many Kinds of Elementary Particles Are There?

How many different kinds of elementary particles are there in the universe? If some physicist succeeds in breaking up an electron next year, what has happened to its claimed elementary nature? More generally, how will we ever know if a particle is truly elementary? Will there

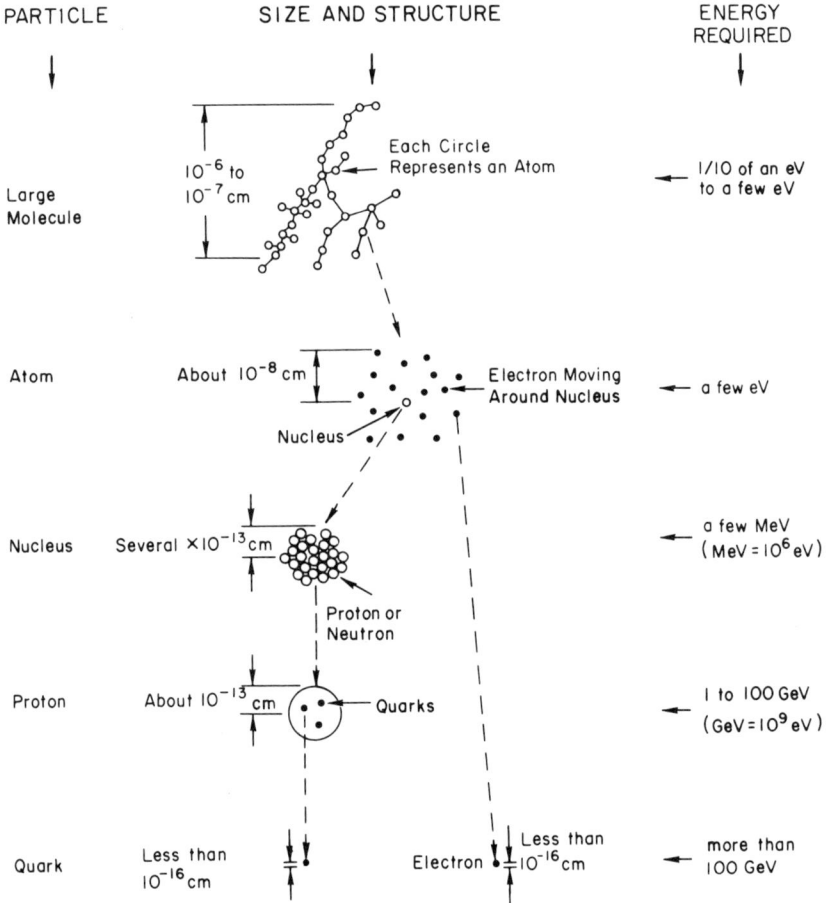

FIGURE 2.1   Many basic objects in nature are made up of yet simpler objects. For example, molecules are made up of atoms, and atoms are made up of electrons moving around a nucleus. To the best of our present knowledge, the elementary particles, electrons and quarks, are not made up of simpler particles. It requires larger energies to investigate the size and structure of the smaller particles. At the right side of the figure are shown the energies required to study the structure of the various objects. The smaller the object, the greater the energy required.

ever be an end to the sequence of particles within particles within particles . . .? In Chapter 3 we describe the present research on these questions. In this section we present a historical perspective.

Figure 2.2 sketches the history of our progress in understanding the number of kinds of elementary particles. The classical Greeks posited just four basic elements: earth, air, fire, and water. In subsequent

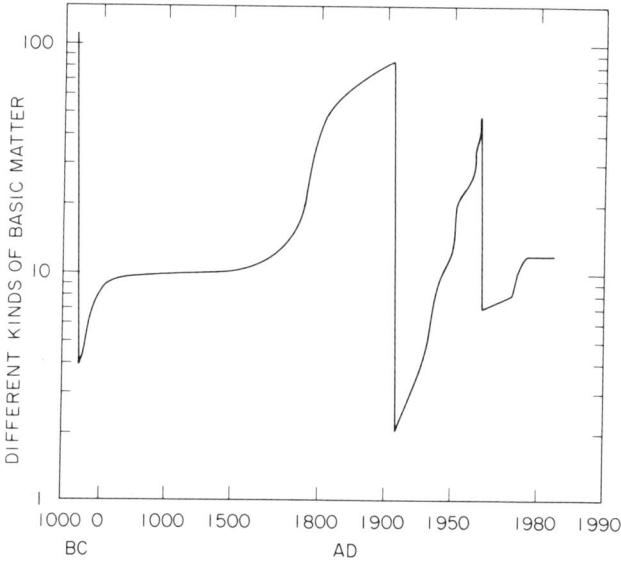

FIGURE 2.2 Mankind has always tried to explain the world as made up of a limited number of different kinds of basic matter. Until a thousand years ago, most people believed that the basic types of matter were earth, air, fire, and water. About 1900 the basic types of matter were thought to be the almost 100 different chemical elements. At present we believe there are about a dozen types of basic matter, namely the leptons and the quarks.

centuries philosophers and alchemists added aether (to include the heavens), mercury, sulfur, salt, and so on. Already we see a simple picture (albeit a wrong one) beginning to expand. In 1661 Boyle defined the concept of a chemical element, and by 1789 Lavoisier had compiled a list of 33 known elements. At this point, a modern particle physicist might have questioned whether these elements were truly elementary. But the list grew steadily, doubling before Mendeleev found a convincing way to classify them into smaller related families in 1868. By 1914 the number of elements had reached 85.

Then revolutionary new developments in physics led to a much simpler picture of matter. Discovery of the electron, the proton, and the tiny dense nucleus of the atom gave rise to the atomic model. Each chemical element consisted of unique atoms, defined by a specific number of electrons surrounding a nucleus made of protons. Thus all matter seemed to be made of only two kinds of constituents, the proton and the electron. A dramatic reduction indeed, from 85 elements to 2 particles.

The neutron was discovered in 1932, providing a more satisfactory picture of the nucleus as a combination of neutrons and protons and increasing the number of fundamental particles to three. In the same year, the positron or antielectron was also discovered. The positron was followed by the muon, the pion, and the first strange particles, all found in cosmic rays. These particles were the first in a long sequence of particles that were unnecessary in the sense that they were not needed as constituents of ordinary matter. Indeed, these particles presented a problem: why did they exist at all, and how were they related to each other? By the 1950s, particle accelerators began to produce hordes of new particles, and their numbers grew in a way quite similar to the number of chemical elements in the nineteenth century (see Figure 2.2).

As before, scientists (now physicists) tried to find patterns in the data that might indicate some underlying simplicity. In 1964 it was proposed that the rapidly growing number of strongly interacting particles (called hadrons) could all be explained as simple combinations of smaller constituents called quarks. There should be three such quarks, and these together with the four known leptons (electron, muon, and their associated neutrinos) would be the seven basic constituents of matter, including the exotic new forms produced only in accelerators. At about the same time, more detailed study of the properties of hadrons, mainly the absence of certain decay processes, caused theorists to suspect the existence of a fourth kind of quark. Such speculation increased with the observation of a new type of force, the weak neutral force. This so-called c or charmed quark was in fact discovered in 1974, as a constituent of a very striking new kind of particle known as the J/ψ. The next year, a new lepton called the τ (tau) was discovered, together with indirect evidence for an associated neutrino ν_τ. In 1976, more charmed particles were discovered, and in 1977 a fifth quark, the b or bottom quark, was discovered.

Thus the number of fundamental constituents of matter has now grown to 11, and if the expected t or top quark is found it will be 12. Is this the final roll call of the elementary particles, or will more be found and the situation once again become complicated? We do not know the answer to that question. Physics, like all the sciences, is based on experimental knowledge. At any given time, all we can do is assemble the full body of our experimental knowledge and try to explain it with a rational and perhaps even elegant theory. If we can explain all of our experimental knowledge with a theory that regards only a certain set of particles as elementary, then that must be sufficient.

The Size of Elementary Particles

As one proceeds down through the sequence of molecule, atom, nucleus, proton, and neutron, and finally quark, the size of the particles gets smaller and smaller. Let us begin with atoms, whose size is of the order of 10^{-8} centimeter (0.00000001 centimeter). This one-hundred millionth of a centimeter is very small by everyday standards. Molecules are larger, their size depending in a rough way on the number of atoms in the molecule. Molecules containing hundreds of atoms, such as organic molecules, can be examined by electron microscopy, and thus can almost be seen in the ordinary sense of that word.

But once we go below the atomic level to nuclei, there is no way to look at these particles with any sort of microscope. The nuclei consist of neutrons and protons packed rather closely together. The proton and neutron are each about 10^{-13} centimeter in size, about 1/100,000 the size of an atom. Nuclei are a few times bigger than a neutron or proton, depending on how many of these particles they contain. But the nuclei are still not much bigger than 10^{-13} centimeter. The sizes of nuclei, neutrons, and protons are too small to be found by looking directly at the particles; they must be measured by indirect methods.

When we come to an elementary particle such as a quark or an electron, we go to a yet smaller scale. By indirect means the sizes of quarks and electrons are known to be less than 10^{-16} centimeter—less than 1/1000 the size of a neutron or proton! Indeed we have no evidence that these particles have any size at all.

Thus the scale of elementary-particle physics is distances of 10^{-13} centimeter and smaller. Elementary-particle physics in its search for the simplest forms of matter has become the physics of the very small.

Elementary Particles and High Energy

At first it seems puzzling that elementary-particle physics, the physics of the very small, is also called high-energy physics. The term high-energy refers to the energies of the particles used to produce particle reactions. By high energy we mean that the kinetic energy (energy of motion) of a particle is much higher than its rest mass energy. Why do we need to carry out our particle reactions with high-energy particles? There are two reasons for this.

First, as Einstein discovered, kinetic energy can be converted into mass, and mass can be converted into kinetic energy. The equation for the conversion is the famous $E = mc^2$, where E is the kinetic energy that can be converted into mass m, and c is the velocity of light. Since we want to produce new particles, and particularly new massive particles, in the reactions that we carry out, we need a large kinetic energy E to make a large mass m.

The second reason for needing high-energy particles is that, as we have already said, we cannot directly see the size of a particle nor directly see if it has internal structure or parts. We must investigate the particle's size and structure by bombarding it with other particles. And the deeper we wish to penetrate into a particle, the higher must be the energy of the bombarding particles.

The famous Heisenberg uncertainty principle also leads to the conclusion that the investigation of small distances requires high energies. If we wish to measure small distances precisely, then there must be a large uncertainty in the momentum associated with that measurement. A large uncertainty in momentum can only be accommodated by a large initial momentum. And large momentum means large energy.

The principal way in which we give high energy to a particle is to accelerate it through an electric field. Thus accelerators are simply machines that have strong electric fields and that guide the particles through those electric fields. (Chapter 5 discusses accelerators and the basic principles of their operation.) This leads to a convenient unit, the electron volt (eV), for measuring both energy and mass. An electron volt is the energy acquired by an electron or proton passing through an electric potential with a total voltage of 1 volt. As we shall see, the electron volt is a rather small unit of energy or mass, so the elementary-particle physicist uses larger units:

$$\text{MeV} = 10^{+6} \text{ eV} = 1 \text{ million electron volts}$$
$$\text{GeV} = 10^{+9} \text{ eV} = 1 \text{ billion electron volts}$$
$$\text{TeV} = 10^{+12} \text{ eV} = 1 \text{ trillion electron volts}$$

The significance of these energy units can be appreciated by looking at some particle masses expressed in electron volts:

1. The electron mass is about 0.5 MeV.
2. The proton mass is about 1 GeV.
3. The heaviest known particle, the Z^0, has a mass of about 100 GeV = 0.1 TeV.
4. New kinds of fundamental particles are predicted by some theories to lie in the still higher mass range of 0.1-2.0 TeV.

In Figure 2.1 we have indicated the range of energies needed to study each type of particle. For the elementary particles shown in the figure, the quark and the electron, the highest energies are needed. In Chapter 5 we describe how the energies of accelerators are related to experimental studies of the elementary particles.

THE KNOWN BASIC FORCES AND FUNDAMENTAL PARTICLES

The Four Basic Forces

One of the great triumphs of physics has been understanding that all the multitudinous phenomena of the material world operate through just four basic forces. We have already mentioned two of these forces: the gravitational and the electromagnetic. Two more were discovered in this century. One is the nuclear or strong force, which holds the nucleus together and also holds the proton and neutron together. The last force to be discovered is called the weak force; we shall describe its behavior below.

Table 2.1 gives some comparative properties of the four forces. The gravitational force is important in our everyday lives and in astronomical phenomena because of the immense mass of the planets and stars. But the gravitational force exerted by one elementary particle is very small compared with the three other forces that can be exerted by that particle.

The electromagnetic forces between elementary particles follow the same laws as the electromagnetic forces that are used in modern technology, such as in motors, generators, and electronic equipment. The elementary particles simply act as small bundles of electric charge and small magnets.

The strongest of the four forces is the nuclear force. However, the nuclear force is not felt directly in everyday phenomena, since it does not extend beyond a distance of about 10^{-13} centimeter from the elementary particle. This distance is about the same as the size of an individual neutron or proton, and thus it determines the size of atomic nuclei. Since atoms and molecules are at least 100,000 times larger, they do not feel the nuclear force. But at distances less than 10^{-13} centimeter the nuclear force is powerful, much more powerful than the electromagnetic force. This is why it is also called the strong force.

Finally we return to the weak force. The distance over which this force acts is also small—less than about 10^{-16} centimeter—and it is much less powerful than the strong force. Yet the weak force is not negligible. In a certain sense it is more pervasive than the strong force. Some elementary particles such as the electron are not affected by the

TABLE 2.1 The Four Basic Forces

Type of Force	Gravitational	Weak	Electro-magnetic	Strong or Nuclear
Behavior over distance	Extends to very large distances	Limited to less than about 10^{-16} cm	Extends to very large distances	Limited to less than about 10^{-13} cm
Strength relative to strong force at a distance of 10^{-13} cm	10^{-38}	10^{-13}	10^{-2}	1
Time for a typical small-mass hadron to decay via these forces		10^{-10} s	10^{-20} s	10^{-23} s
Particle that carries the force	Not discovered	W^+, W^-, and Z^0; intermediate bosons	Photon	Gluon. The gluon has been identified indirectly but it has not been, and perhaps cannot be, isolated.
Mass of particle	Not known	About 90 GeV	0	Assumed 0

strong force but are affected by the weak force. The radioactive decay of the neutron and of nuclei, as well as the decays of many of the elementary particles, occur through the weak force.

Since the 1920s physicists have speculated about the possibility that different forces can be unified into one general theory. That is, are the seemingly different forces simply different manifestations of one general force? First thoughts were about unifying the gravitational and electromagnetic forces; that has not been done, and we do not know if it can be done. But within the last 15 years, a significant unification of the electromagnetic and weak forces has been made and has been verified experimentally. In Chapter 3 the state of current research on force unification is discussed.

The Known Families of Elementary Particles

At present, all our observations in particle physics can be explained by the existence of the four basic forces and by the existence of three families of elementary particles. These families are the leptons, the quarks, and the force-carrying particles.

THE FORCE-CARRYING PARTICLES

We turn first to this family of elementary particles. It is a basic principle of quantum mechanics that a force has a dual nature: it can be transmitted through a wave or through a particle. The clearest example is the electromagnetic force, which can be treated in some situations as being carried by an electromagnetic wave (radio waves or light waves, for example) and in other situations as being carried by a particle (the photon). The question then arises whether the other forces also obey quantum mechanics in this sense and thus can be thought of as being carried by particles. Table 2.1 summarizes our present knowledge. The weak force is indeed carried by particles: the W^+, W^-, and Z^0 intermediate bosons have recently been discovered. We believe that the strong force is also carried by particles called gluons, but here the evidence is indirect. Unlike the photon, W^+, W^-, and Z^0, the gluon has not been isolated. Finally, the particle conjectured to carry the gravitational force has been called the graviton, but such a particle has not yet been discovered, and there is no experimental evidence for its existence. Because of the feebleness of the gravitational interaction among elementary particles, its detection would be extraordinarily difficult.

THE LEPTONS

The lepton family of elementary particles is defined by two properties:

1. Leptons are affected by the gravitational, electromagnetic, and weak forces but not by the strong force.

2. Leptons must be either created or destroyed in particle-antiparticle pairs; the total number of leptons (number of leptons minus number of antileptons) is conserved in all processes to the best of our knowledge.

Figure 2.3 shows the six known leptons. They come in pairs, each pair consisting of one charged lepton and one neutral lepton. The neutral lepton is called a neutrino. Each pair is called a generation, and in each generation the mass of the neutrino is much less than the mass of the charged lepton.

In the last few years there has been speculation, but as yet no evidence, that the proton might very rarely decay to a lepton plus hadrons. If that turns out to be true, the total number of leptons would not be conserved in this process.

Generation	Particle	Charge	Mass

| 1 | electron (e) | -1 | 0.51 MeV |
| | electron neutrino (ν_e) | 0 | less than 50 eV |

| 2 | muon (μ) | -1 | 106 MeV = 0.106 GeV |
| | muon neutrino (ν_μ) | 0 | less than 0.5 MeV |

| 3 | tau (τ) | -1 | 1784 MeV = 1.784 GeV |
| | tau neutrino* (ν_τ) | 0 | less than 160 MeV = 0.160 GeV |

*indirect evidence

FIGURE 2.3 The six known leptons are arranged in pairs. The members of a pair interact only with each other. For example, the electron and electron neutrino interact with each other but not with the muon, the muon neutrino, the tau, or the tau neutrino. There is indirect evidence for the tau neutrino; it has not been directly detected.

The questions that we now face are profound. Are there more generations of leptons? What sets the mass of the leptons, and the difference in masses between generations? And of course the ultimate question: are the leptons really elementary?

THE QUARKS

The quark family of elementary particles (Figure 2.4) is also defined by two properties:

1. Quarks are affected by all four basic forces. Because they are affected by the strong force, quarks act very differently from the leptons in many situations. In particular, it is either impossible or very difficult to isolate quarks, whereas leptons can easily be isolated.

2. Quarks, like leptons, cannot be singly created or destroyed to the best of our knowledge. Therefore the number of quarks, like the number of leptons, is conserved in every physical process.

A very peculiar property of the quarks is that they have electric charges of 2/3 or 1/3 of the unit of electric charge carried by the electron and the proton. All other particles, elementary or not, have either zero or integral charges. Like the leptons, the quarks fall into pairs called generations. Each pair has a +2/3 unit charge quark and a −1/3 unit charge quark.

Five quarks have been discovered. Most particle physicists believe that there is a sixth quark, called the t or top quark, which will complete the third generation.

There is an important unanswered question concerning quarks. Can a single quark be isolated from all other matter so that it exists all by itself as a free particle? We know from experiment that all the leptons can exist as free particles. But can the quarks be free? At present most physicists believe that quarks are always confined in more complicated particles such as protons. This belief is based on the failure of almost all experiments that have tried to make or find free quarks. We say almost all because there has been a series of experiments that have indicated that free quarks might exist. In the end this is an experimental question, and the search for its answer is one of many reasons for wanting to do experiments at very high energies.

THE HADRONS

Before concluding this section, we briefly describe the vast hadron family of particles. Hadrons are subnuclear particles, but they are not elementary particles. To the best of our knowledge, hadrons are made of either three quarks or one quark and one antiquark bound together by the strong force. Table 2.2 lists a few of the known hadrons. The first hadrons to be discovered were the proton and neutron. Now more than a hundred types of hadrons are known.

Generation	Particle	Charge	Mass
1	up (u)	+2/3	about 300 MeV = 0.3 GeV
	down (d)	−1/3	about 300 MeV = 0.3 GeV
2	charm (c)	+2/3	about 1500 MeV = 1.5 GeV
	strange (s)	−1/3	about 500 MeV = 0.5 GeV
3	bottom (b)	−1/3	about 5,000 MeV = 5.0 GeV

FIGURE 2.4   Five quarks are well known. The up and down quarks form one pair; the charm and strange quark form a second pair. There are strong theoretical reasons for assuming the existence of a sixth quark, called the top quark, to be paired with the bottom quark. As of mid-1984, there is initial experimental evidence for the existence of the top quark. Unlike the leptons there can be interactions between the quark pairs. For example, a strange quark can decay to an up quark.

TABLE 2.2   Some Hadrons with Their Masses and the Quarks
They Are Made of

Name	Symbol	Mass in GeV	Quarks in the Hadrons
Proton	p	0.938	2 up quarks plus 1 down quark
Antiproton	$\bar{p}$	0.938	2 anti-up quarks plus 1 anti-down quark
Neutron	n	0.940	1 up quark plus 2 down quarks
Positive pion	π^+	0.140	1 up quark plus 1 anti-down quark
Positive kaon	K^+	0.494	1 up quark plus 1 anti-strange quark
J or psi	J/ψ	3.097	1 charm quark plus 1 anti-charm quark
Upsilon	Υ	9.460	1 bottom quark plus 1 anti-bottom quark

Although hadrons are not in themselves elementary particles, they are nevertheless important in elementary-particle physics research. First, we do not know how to isolate quarks, so to do experiments on quarks we must use the quarks in hadrons. Second, hadrons are a fascinating form of matter, and it is interesting to study them in their own right.

The strong force, which holds the quarks together in the hadron, is carried by gluon particles as described earlier. It is useful to think of the gluons as traveling between the quarks, being emitted by one quark and absorbed by another quark. Thus the hadron may be thought of as being composed of gluons as well as quarks. Indeed in a moving hadron the gluons carry part of the energy. However, it is the quarks that determine the mass and other properties of the hadron.

PARTICLES AND ANTIPARTICLES

In Figures 2.3 and 2.4 we listed the six known leptons and five known quarks. For each of the particles listed there exists a related particle, of the same mass but opposite electric charge, called its antiparticle. Thus the electron, which has negative charge, has a related particle called the antielectron or positron, which has positive charge. This same relation applies to the quarks. For example, the bottom quark has a charge of $-1/3$; the bottom antiquark has the same mass but has a charge of $+1/3$. Hadrons as well have their antiparticles. The most famous example is the antiproton, the antiparticle of the proton. Indeed, relativistic quantum theory dictates that every particle must have an antiparticle.

As noted earlier, single quarks and single leptons cannot be either created or destroyed. However, it is possible in a collision for a particle and its antiparticle to annihilate to form energy (in the form of a photon of light, for example). Similarly, it is possible to produce a new particle-antiparticle pair in a collision of two high-energy particles, by converting some of the collision energy into the mass of the particle-antiparticle pair. Some examples of these processes are discussed in the next section.

COLLISIONS AND DECAYS

Collisions of Particles

Elementary particles and hadronic particles are too small to be studied directly. We study them indirectly by colliding two particles together and then determining what particles come out of the collision. Each time there is a collision, a number of different things can happen. This is illustrated in Figure 2.5, which shows two protons colliding head-on in a proton-proton colliding beam accelerator. Figure 2.5 shows a time sequence: in (a) the protons are about to collide, and in (b) they have just collided to form a complicated concentration of mass and energy. This concentration of mass and energy is unstable, and it can change again into particles in many different ways. Thus in (c) two protons may come out again, or a large number of hadrons may come out of the collision, or other kinds of particles not shown here may be produced. There are many possibilities. By studying all the different particles that come out of collisions, we do two things. We learn about the forces between particles, and we can also find new kinds of particles.

Collision Diagrams

It is clumsy to draw the time sequence of several diagrams such as those shown in Figure 2.5. Instead, physicists use a kind of shorthand

(a) Protons about to collide head-on.

(b) Concentration of mass and energy just after protons collide.

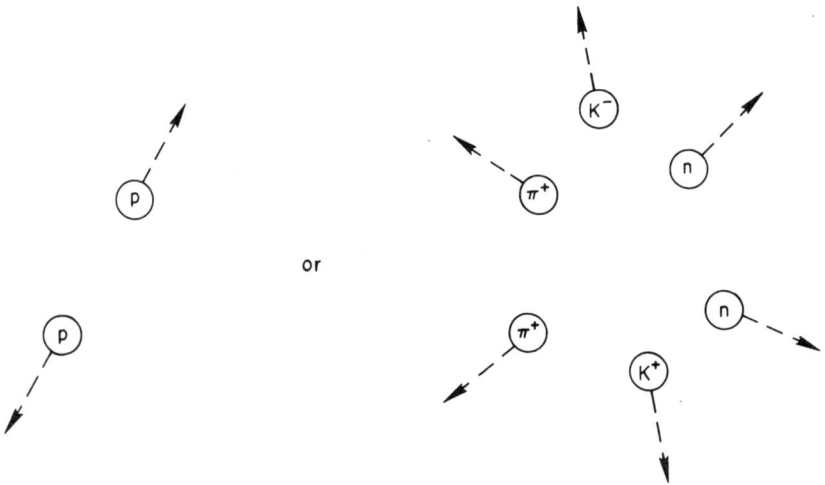

or

(c) Two possibilities for what may come out of the collision.

FIGURE 2.5   In (a) two protons are about to collide head-on. When they collide as in (b), their mass and energy are concentrated in a small region of space. That concentration of mass and energy is unstable and very quickly breaks up into new particles as in (c). Sometimes just two particles come out of the collision. But at high energies usually many particles come out of the collisions, and none of them needs to be the original protons.

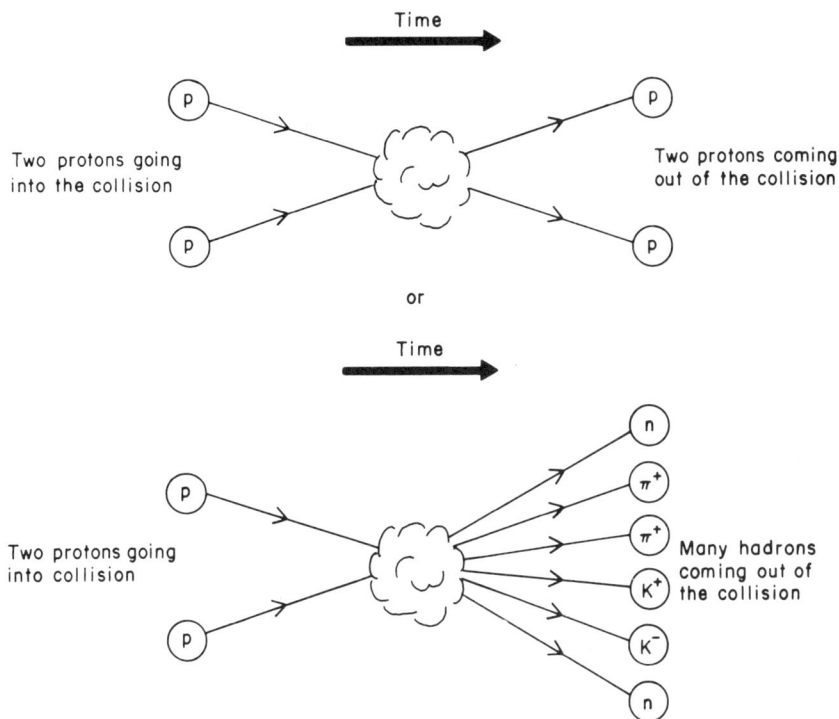

FIGURE 2.6 The collisions of particles can be represented succinctly in a diagram in which time advances from left to right. For example, the lower figure shows two protons going into a collision producing a concentration of mass and energy, which then breaks up into six particles.

in which the collision process is pictured in a single diagram. These pictures are of great assistance in making calculations, and in this context they are called Feynman diagrams after their inventor. Thus Figure 2.6 shows the collision diagram for two protons going into two protons, and also the process for two protons going into many hadrons—the same two processes shown in Figure 2.5. (Note that we use a convention in which time advances from left to right.)

Collisions and Interactions

The concentration of mass and energy in Figures 2.5 and 2.6 represents the crux of how particles interact through the basic forces. Often we know enough about that concentration region to explain it in simple terms. For example, when an electron and positron collide they

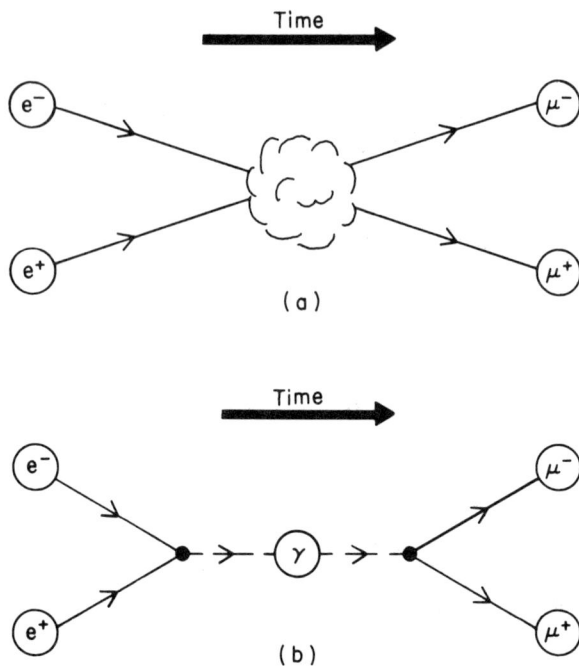

FIGURE 2.7   The collision of an electron and a positron can lead to the production of a positive muon and a negative muon. The electron and positron actually disappear; the technical term is that they annihilate each other. A sketch of how that interaction occurs is shown in (a). A more detailed description of the interaction is given in (b), called a Feynman diagram. The electron and positron annihilate to produce a photon, the particle that carries the electromagnetic force. The photon carries the unstable concentration of mass and energy and quickly changes into the pair of muons.

can make a muon (μ^-) and an antimuon (μ^+). We know how this occurs and can draw the collision diagram as shown in Figure 2.7. As time advances (i.e., moving to the right in the figure), the electron and positron collide; the collision annihilates the electron and positron and produces a highly excited photon, which contains all the collision energy. This photon is extremely unstable and quickly decays into a muon and an antimuon. The photon could alternatively produce a quark-antiquark pair or another electron-positron pair.

Particles can also interact by exchanging a force particle. For example, an electron can scatter off a muon by emitting a photon that is absorbed by the muon, as shown in Figure 2.8. The photon carries energy and momentum from the electron to the muon, causing both particles to deflect.

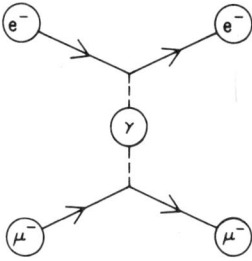

FIGURE 2.8 This Feynman diagram shows how an electron scatters off a muon. Note that the electron and muon do not interact directly, but rather through the exchange of a photon (γ).

In general, interactions between two particles can be represented by one of these two types of diagrams or by somewhat more complicated diagrams in which more than one force particle is exchanged.

Spontaneous Disintegration of Particles

Many elementary particles and almost all hadrons are unstable, even if isolated from any external forces. Eventually they decay to particles of smaller mass, which in general were not components of the original unstable particles. For example, the hadron called the pion decays into a muon plus a neutrino. The average length of time that a particle survives before it decays is called its lifetime. The lifetimes of most of the known particles are very short by everyday standards, ranging from 10^{-6} to 10^{-22} second.

Indeed, few completely stable particles are known. The stable ones appear to be the proton, the electron, and the neutrinos. But the stability of even the proton has recently been called into question, as described in Chapter 4.

CONSERVATION LAWS AND SYMMETRY IDEAS

What Are Conservation Laws?

As particles interact and decay, they present a picture of a world dominated by change. To bring order to this world, the physicist looks for properties of matter that do not change.

A simple example of an unchanging quantity is the total energy of the particles in a collision. If the masses are counted as a part of the total energy, using $E = mc^2$, then no matter how the particles collide or what comes out of the collision, the total energy is unchanged. We say that the total energy is conserved or equivalently that there is a conservation law for the total energy. Another example is that the total electric charge never changes in an interaction; hence there is a conservation law for the total electric charge.

We have seen that to the best of our knowledge single leptons and quarks cannot be either created or destroyed, but that particle-antiparticle pairs can be produced and can annihilate. These observations are expressed in the form of two new and important conservation laws: the law of lepton number conservation and the law of quark number conservation.

To apply these laws, we assign a positive lepton number of $+1$ to each lepton and a negative lepton number of -1 to each antilepton. Then the total lepton number in any interaction, obtained by adding the lepton numbers of each particle, can never change. Lepton-antilepton pairs can still be produced, however, since the total lepton number changes by $(+1) + (-1) = 0$; thus lepton number is conserved in such a process. Similarly, quarks carry a quark number of $+1$ and anti-quarks carry a quark number of -1, so the total quark number is always conserved. Force particles do not carry either quark or lepton numbers, and there is no conservation law for force particles. Thus it is possible to create a single photon.

Conservation laws such as these have deep significance for our understanding of the basic nature of matter. But in all cases they are based not on philosophy but on experiment. Indeed, new experiments may find instances in which such laws are not obeyed. Then the laws will have to be modified or dropped altogether.

Symmetry and Invariance

Conservation laws provide one sort of regularity and certainty in the world of interacting and decaying particles. Another sort of regularity is provided by symmetry ideas. Symmetry here is an extension of the idea of symmetry in patterns and designs.

In constructing new physical theories it is helpful to be guided by considerations of symmetry. The aesthetic appeal that symmetry has held for many cultures is evidenced by the pleasure we find in regular decorative patterns in natural forms such as crystals.

Symmetry may be understood as a motion that leaves the form of a pattern or an object unchanged in appearance. For example, the four-bladed windmill, Figure 2.9, possesses a fourfold symmetry. After rotation by 90 degrees about its axis it looks identical to its unrotated self. A sphere is invariant after any rotation about its center; in other

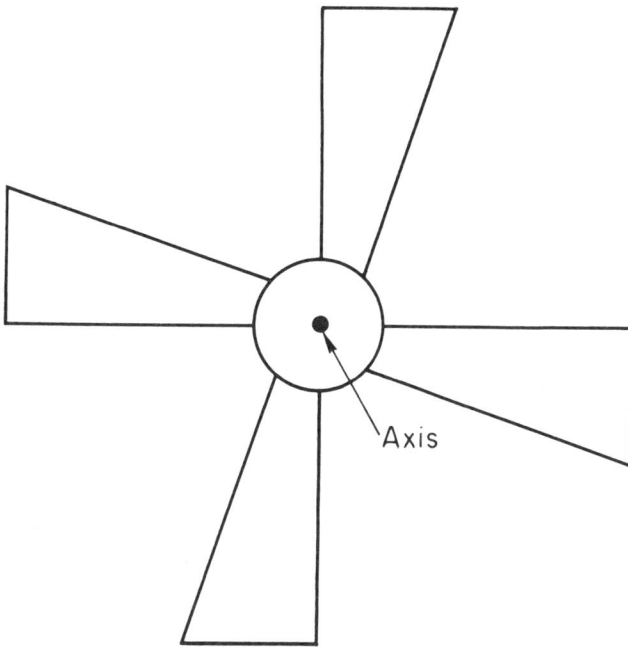

FIGURE 2.9   An example of fourfold symmetry in a four-bladed windmill. Its axis of rotation, marked by the central black dot, is perpendicular to the paper. The picture is not changed by a 90° rotation about that axis.

words, it looks the same from all sides. Invariant, a word that occurs frequently in physics, means unchanged. Physical theories can have symmetries of a similar kind, but what remains invariant or unchanged after a transformation is not a pattern or an object but the mathematical structure of the laws of the theory itself. Physicists now agree that symmetries play a central role in our understanding of nature.

The twin concepts of symmetry and invariance can be important in limiting the equations and theories that are applied to a phenomenon. Consider the force of the Earth's gravity on a person walking on the Earth's surface, and use the good approximation that the Earth is a sphere. Then without knowing anything about the laws of gravitational force, we can make two statements from just the arguments that a sphere is symmetric about its center for any rotation and that the gravitational force must be invariant to any such rotation. First, the size of the force must be the same, no matter where the person walks

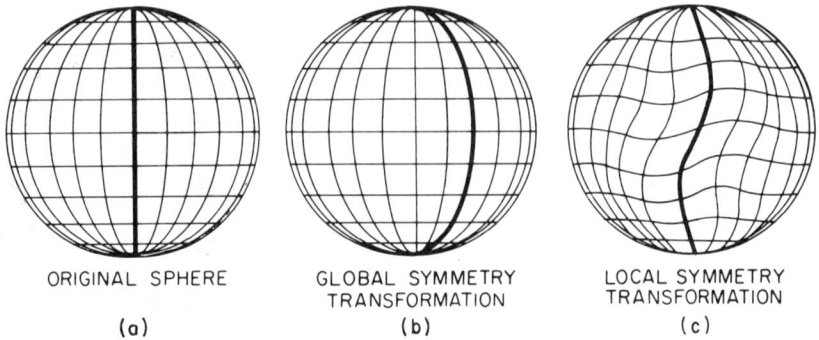

ORIGINAL SPHERE   GLOBAL SYMMETRY   LOCAL SYMMETRY
                          TRANSFORMATION   TRANSFORMATION
      (a)   (b)   (c)

FIGURE 2.10   The ideas of global and local symmetry can be illustrated by a sphere marked with lines of longitude and latitude. When the sphere is simply rotated about its axis the shapes of the lines are not changed; that is called a global symmetry transformation. If the surface of the sphere is distorted as one might do with a sphere made out of rubber, such that the lines of longitude and latitude are twisted, that is a local symmetry transformation.

on the Earth. Second, the force must point directly toward the Earth's center or directly away. It cannot point in any other direction, east for example, since that direction is not invariant to a rotation. But this is as far as this symmetry argument can go; it cannot tell us whether the force is up or down or its strength. To know that, we need first experiment and observation then a theory with explicit equations.

Physicists use other symmetry and invariance ideas in much the same way, to provide some general information and to limit the range of equations and theories that can apply. This is particularly important in particle physics where the basic objects, the elementary particles, are relatively simple and have many kinds of symmetries.

The symmetries of physical theories are of two types, called global and local. The distinction between them may be illustrated by considering an ideal spherical balloon [Figure 2.10(a)] marked with a system of latitude and longitude coordinates so that the positions of all points on the surface can be identified. A global symmetry is exhibited if the sphere is rotated about some axis [Figure 2.10(b)]. In geographical terms, the rotation depicted is equivalent to displacing the prime meridian from Greenwich, England, to Alexandria, Egypt. This rotation is a symmetry operation because the form of the sphere remains unchanged. It is called a global symmetry because the locations of all the points on the surface are changed by the same angular displacement in longitude.

Local symmetry is a more demanding statement. It requires that the balloon maintain its shape even if the points on the surface are displaced independently [Figure 2.10(c)]. A local symmetry operation stretches the balloon and therefore introduces forces between points. Each of the fundamental forces is now thought to arise from a similar requirement that a law of Nature be invariant under local symmetry transformation. Because the earliest attempts to construct interactions from symmetries dealt with invariance under a change of scale or gauge, the resulting theories are called gauge theories.

The symmetries we have discussed so far are known as continuous symmetries, because they may be built up from infinitesimal motions. Another important class of symmetries of physical laws is made up of discrete, or discontinuous, transformations. Of these the most familiar in everyday experience is left-right or mirror symmetry, which is manifested by many objects in our environment. Many microscopic physical processes are invariant under time reversal; a film of the event, run backwards, would also correspond to an allowable event. Similarly, in many situations the replacement of all particles by their antiparticles leads to no change in the physical outcome. As an illustration, the light emitted by an antineon lamp would be indistinguishable from the light emitted by a conventional neon lamp.

Symmetry Breaking

It may happen that the laws of physics embody a certain symmetry, but some of their consequences do not manifest that symmetry. An example will show how this may come about. Above a certain critical temperature, the individual microscopic magnets that make up an iron ferromagnet are oriented randomly. This reflects the invariance of the laws of electromagnetism under rotations, which is to say that there is no preferred direction in space. When the iron is cooled below the critical temperature, the micromagnets tend to align themselves along some randomly chosen direction. The randomness of this direction is attributable to the rotational invariance of electromagnetism. Once the micromagnets have frozen along a certain direction, the ferromagnet does not display rotational invariance, because a specific direction has been singled out. Thus the symmetry of the laws of electromagnetism has been hidden.

In elementary-particle physics, the most striking case of symmetry hiding occurs in the theory of weak and electromagnetic interactions. There the equations of the quantum theory possess a local gauge symmetry, but the observed particles such as electrons do not display this symmetry in their masses.

EXPERIMENTS, ACCELERATORS, AND PARTICLE DETECTORS

Experimental Methods in Elementary-Particle Physics

The purpose of experiments in elementary-particle physics is to study the behavior of the forces that act on the particles and to look for new types of particles and forces. But few of these studies and searches can be carried out using the apparatus found in the usual physics laboratory. For example, elementary particles are too small to be seen using a visible light microscope or even an electron microscope. Furthermore, many elementary particles have short lifetimes; they simply do not exist for a long enough time to be studied directly. A final example is that the search for new particles usually requires that other particles collide together at high energies to produce the new particles.

The primary experimental method in elementary-particle physics involves the collision of two particles at high energy and the subsequent study of the particles that come out of such a collision. We are interested in the kinds of particles that come out of the collisions, how many there are, the energies of the particles, and their directions of motion. In this section we give an overview of how such experiments are done.

Experiments at Fixed-Target Accelerators

The basic concept of an elementary-particle experiment using an accelerator is shown in Figure 2.11. A beam of protons is accelerated to high energy by a proton accelerator. The beam of protons leaves the accelerator and passes into a mass of material called a target, which is fixed in position. The collisions occur between the protons in the beam and the material in the target. Hence this is called a fixed-target accelerator, and the experiment is called a fixed-target experiment.

The simplest material to use for the target is hydrogen, because the hydrogen atom consists of a single electron moving around the single proton that forms the nucleus of the hydrogen atom. Most of the time the protons in the high-energy beam will pass right through the hydrogen target without striking anything, but occasionally one of the protons in the beam will hit either a proton or an electron in the hydrogen. We restrict our attention here to the case when a proton in the beam hits a proton in the hydrogen atom. Then we have a proton-proton collision. As discussed earlier in this chapter in the section on Collisions and Decays and sketched in Figure 2.6, one of the

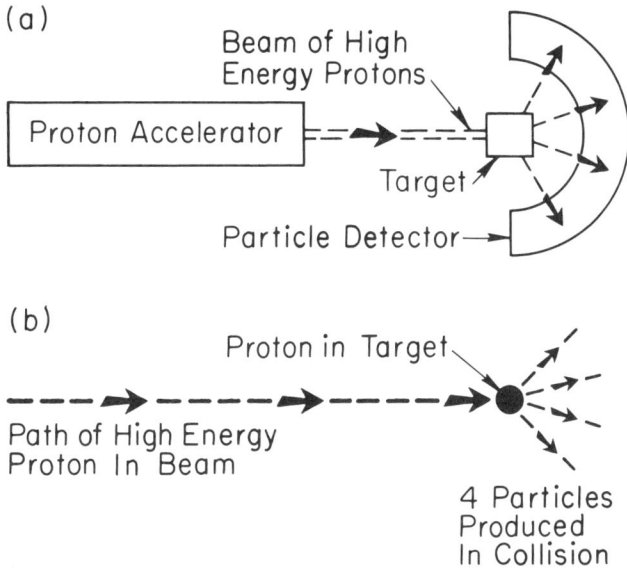

FIGURE 2.11   In a fixed-target experiment a beam of high-energy particles, for example protons, is produced by an accelerator. The beam of particles interacts with the target producing new particles. The particles are detected and their properties studied using an apparatus called a particle detector. In (a) the entire experiment is sketched. In (b) the interaction of the particle itself is shown: a proton in the beam interacts with a proton in the target and produces four particles.

things that can happen is that two protons can simply come out of the collision again. But sometimes many other particles—hadrons and leptons—can come out of the proton-proton collision.

In order to determine what has happened, we need an apparatus that can detect the particles coming out of the collision. Such an apparatus is called a particle detector (see Figure 2.11). Particle detectors cannot see particles directly, but they can determine their energies and directions of motion and the nature of the particles. How this is done is described below. Thus the three basic elements of experiments at fixed-target accelerators are the accelerator, the target, and the particle detector. We next describe each of these elements in more detail.

Fixed-Target Accelerators

The particles accelerated must be stable and have electric charge, hence either protons or electrons are used. The acceleration process begins with these particles at rest, and gradually gives them more and more energy until they are moving with speeds close to the speed of light and have high energy. The particles are given the energy by the force of electric fields acting on their charge. Since there is a limit to how strong an electric field we can make, higher energies require larger accelerators.

High-energy accelerators are large and expensive machines. Thus few are built, and these are used as intensively as possible. For example, in the United States there are only two high-energy proton accelerators. The Alternating Gradient Synchrotron (AGS) at Brookhaven National Laboratory has a maximum energy of about 30 GeV and has been in operation since 1960. The Tevatron at Fermi National Laboratory, a circular accelerator with a diameter of 2 kilometers, has just gone into operation; it is the first large accelerator in the world to use superconducting magnets, and it is designed to reach an energy of 1000 GeV. Also in the United States is the 3-kilometer-long electron accelerator at the Stanford Linear Accelerator Center. (The complementary uses of the different energy ranges and particle beams are described in Chapter 5.) In addition, the United States has lower-energy proton and electron accelerators that are used primarily for nuclear-physics research.

Targets

We have already described how hydrogen can be used as a target for the beam of particles coming out of an accelerator. Other materials can also be used as targets. For example, deuterium is often used. In deuterium (heavy hydrogen) the nucleus consists of a proton plus a neutron; hence one can study collisions between the protons or electrons coming out of the accelerator and the neutron in the target. Another example is provided by neutrino experiments, which often require a dense target such as iron.

Particle Detectors for Charged Particles

Not only charged particles, such as protons or charged pions or electrons, but also neutral particles, such as neutrons and photons, can come out of a collision. Charged means that the particle has positive or negative electrical charge, as opposed to a neutron or photon, which have no electrical charge. No particle can be seen directly, but as a charged particle passes through any kind of material, it breaks up the atoms and molecules in that material. The technical term is that it ionizes the material. And through that ionization the path of the charged particle can be determined.

The bubble chamber provides the classic example. The liquid in a bubble chamber is heated above its boiling point, but it is prevented from boiling by high pressure in the chamber. If that pressure is released for a short time and then reapplied, the liquid still does not boil. However, if a charged particle passes through the chamber while the pressure is released, the resulting ionization leads to the formation of a string of bubbles along the path of a particle. This string of bubbles can be photographed, as shown in Figure 2.12, to produce a picture of the tracks or paths taken by the charged particles in their passage through the chamber.

Ionization produced by a charged particle is used in other ways by other types of particle detectors. In a drift chamber, for example, the charged particle ionizes a gas, and the electrical effect of that ionization is used to determine the particle path. In a scintillator, the ionization produces visible light that is detected by a phototube. Some particle detectors, such as Cerenkov radiation detectors, do not use ionization. Chapter 6 describes particle detectors in detail, including a discussion of how neutral particles are detected.

Secondary Particle Beams

The primary beam produced in an accelerator is always either protons or electrons, because stable and charged particles must be used for the acceleration process. Once the primary beam of protons or electrons leaves the accelerator, it is often used to produce secondary beams of other kinds of particles. Figure 2.13 provides an example in which the primary proton beam from a proton accelerator is used to produce a secondary beam of charged pions. This is done in a production target in which the protons interact with the target material to produce the pions. The beam of pions then passes into a bubble chamber; in this example the chamber liquid is hydrogen. The pions finally interact with the electrons and protons in the hydrogen, those being the collisions that are being studied. Other examples of secondary particle beams are neutrino beams, muon beams, and photon beams.

Particle Colliders

In many elementary-particle physics experiments it is important to have very-high-energy collisions. Therefore through the years acceler-

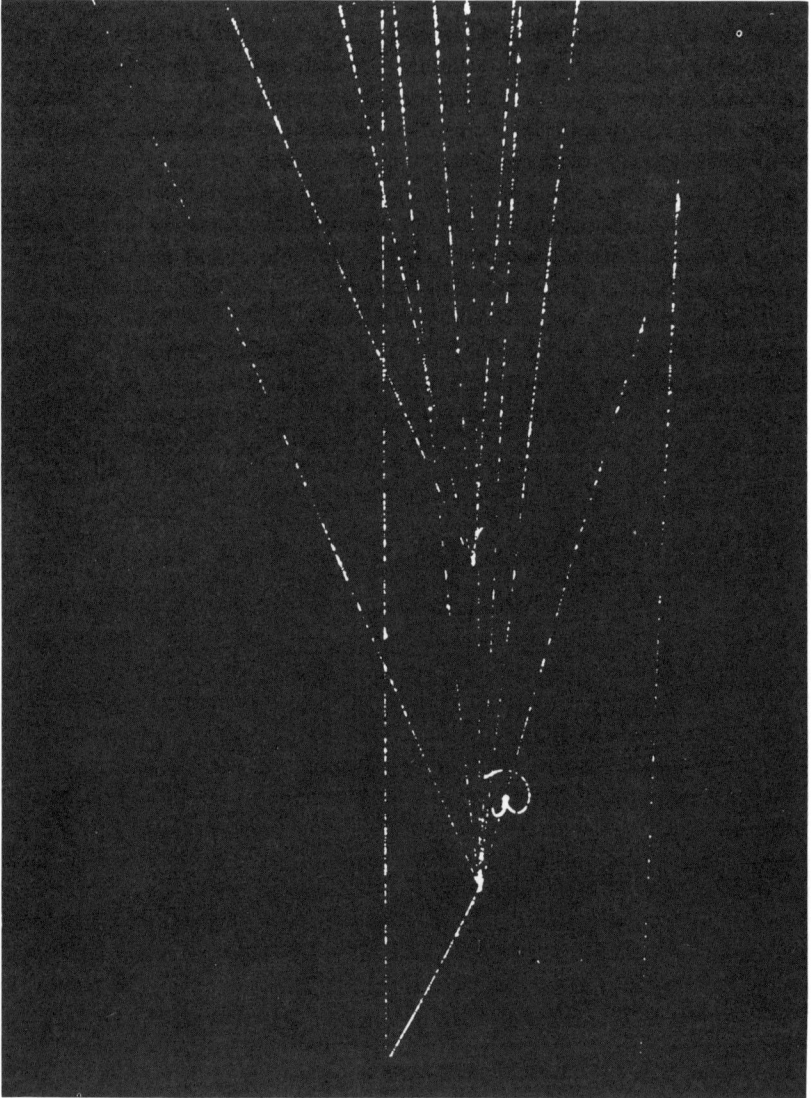

FIGURE 2.12 An example of a photograph of charged-particle tracks in a bubble chamber. Two sprays of particles emerge from the two vertex points at which they were created. The upper vertex is the point at which a neutral charmed meson decayed into four charged particles: $D^0 \rightarrow K^+\pi^+\pi^-\pi^-$. The decay distance was 9 millimeters, which corresponds to an unusually long lifetime for this particle of 5.5×10^{-12} second. The photograph is from the SLAC Hybrid Facility Photon Collaboration.

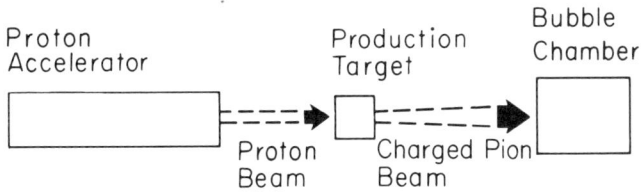

FIGURE 2.13   In many accelerator experiments the primary particle beam from the accelerator is used to produce a secondary beam, and experiments are carried out with the secondary beam. For example, a proton accelerator can be used to produce a beam of charged pions through the interaction of its primary beam with a production target. The secondary beam of pions is then used for experiments.

ator builders have put higher and higher energy accelerators into operation: our phrase is "pushing the energy frontier." But in fixed-target experiments the useful energy for the collision does not increase nearly so fast as the energy of the primary beam increases. Hence in fixed-target accelerators it becomes increasingly expensive to keep pushing the energy frontier.

The alternative is to collide two beams of particles moving in opposite directions, as shown in Figure 2.14. In this case the useful energy is actually the sum of the energy of each of the two beams (if the two beam energies are equal). Particle colliders now produce the highest useful energy of any of our machines.

In particle colliders both beams must consist of stable, charged particles; the choice in practice has been restricted to protons and electrons and to their antiparticles—antiprotons and positrons. The most common form of collider uses opposing beams of electrons and positrons. This is because the collision of an electron and a positron is often relatively simple to understand. On the other hand, the highest-energy collisions are at present obtained with protons colliding with antiprotons.

In Chapter 5, the section titled Accelerators We Are Using and Building describes the world's particle colliders; here we give a few examples. Operating electron-positron colliders range in energy from a few GeV to 45 GeV. The Stanford Linear Collider under construction in the United States will yield 100 to 140 GeV in energy, and the LEP electron-positron collider being constructed at the CERN laboratory in Europe can eventually reach over 200 GeV. CERN is now operating a proton-antiproton collider with a total energy of over 500 GeV, and the Fermi National Accelerator Laboratory in the United States has a 2000-GeV proton-antiproton collider under construction. The elementary-particle physics community in the United States is now discussing

(a)

Collision
Occurs Here

Target

FIXED TARGET

(b)

Collision
Occurs Here

COLLIDING BEAMS

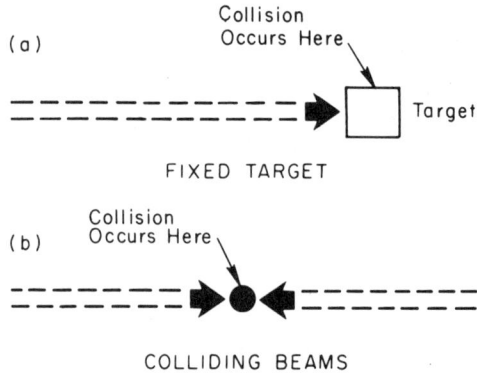

FIGURE 2.14   (a) In fixed-target experiments, a beam of high-energy particles collides with particles at rest in a target. (b) In colliding-beam experiments, two beams of high-energy particles collide head-on. Colliding-beam experiments allow the experimenter to reach much higher effective energies when studying the interactions of particles.

the possibility of the construction of a proton-proton collider to reach 40,000 GeV.

Experiments at Particle Colliders

Since there is no fixed target in a particle collider, the particle detector must look directly at the region where the opposing beams of particles collide. Figure 2.15 shows how this is done in a circular collider where the beams of particles move in opposite directions

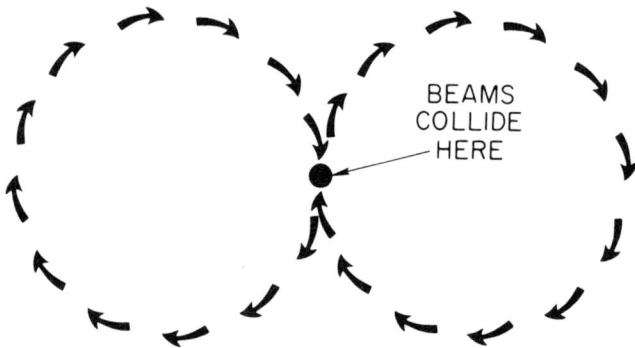

BEAMS
COLLIDE
HERE

FIGURE 2.15   In the simplest form of colliding-beam facilities, two beams of particles rotate in the same direction in circles that are tangent at just one point. The beams collide at that point.

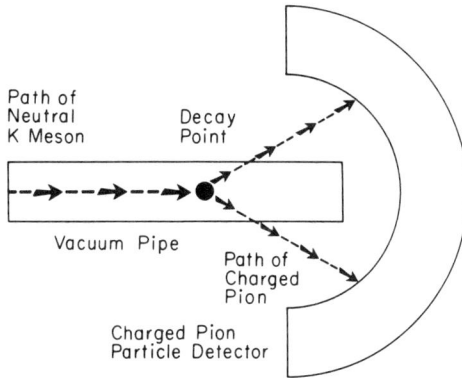

FIGURE 2.16   Sometimes the decay of a particle is of interest. The sketch shows how the decay of a neutral *K* meson into two charged pions is studied. This is one of the crucial experiments in the study of CP violation.

around two circles. In this simple example the beams collide at just one point. In a real collider, the beams would be arranged to collide at several different points, providing the opportunity to carry out several experiments at once.

The Decays of Particles

Until now we have discussed the most common form of experiment in which the collision of two particles is studied. Sometimes, however, we study the decay of a single particle. Figure 2.16 illustrates this by an experiment that studies the decay of a neutral *K* meson to two charged pions.

Experiments in Elementary-Particle Physics Without Accelerators

A large variety of experiments in elementary-particle physics is carried out without using accelerators. Some of the experiments use particles from fission reactors or from cosmic rays. Others look for new particles, such as free quarks or magnetic monopoles, in ordinary matter. Still others study with great precision the properties of the stable or almost stable particles, testing, for example, the equality of the size of the electric charge of the electron and the proton. In Chapter 6, the section on Facilities and Detectors for Experiments Not Using Accelerators takes up this subject.

3

What We Have Learned in the Past Two Decades

DEVELOPMENT OF THE QUARK MODEL OF HADRONS

The Beginnings of the Quark Model

It was first recognized in 1964 that all the known hadrons fell into the particular symmetry scheme, or pattern, expected if all hadrons are formed from three fundamental constituents. These were called quarks, and they were given the names u, d, and s for up, down, and strange. Each hadron would be composed of either three quarks [such as the ten-member group shown in Figure 3.1(A)] or of quark-antiquark pairs [such as the octet group shown in Figure 3.1(B)].

Note that for each of these states the total charge of the particle is the sum of the charges of the quarks of which it is composed. For example, the Δ^{++} (pronounced delta plus plus) shown in Figure 3.1(A) consists of 3 u quarks, so its total charge is $3 \times (+2/3) = +2$ (hence the $++$ superscript). Similarly, the Δ^{+} is composed of uud and has a charge of $2 \times (+2/3) + (-1/3) = +1$.

Each of these quarks is in a particular orbit or state of motion relative to the other two quarks. If we were able to reach into the Δ^{++} and magically transform one of the u quarks into a d quark, without altering the orbit of the quark, then we would have a Δ^{+} (delta plus) particle. Similarly, if we were to change one of the two u quarks in the Δ^{+} into a d quark, then we would have a Δ^{0} (delta zero) and so on. The similar

48

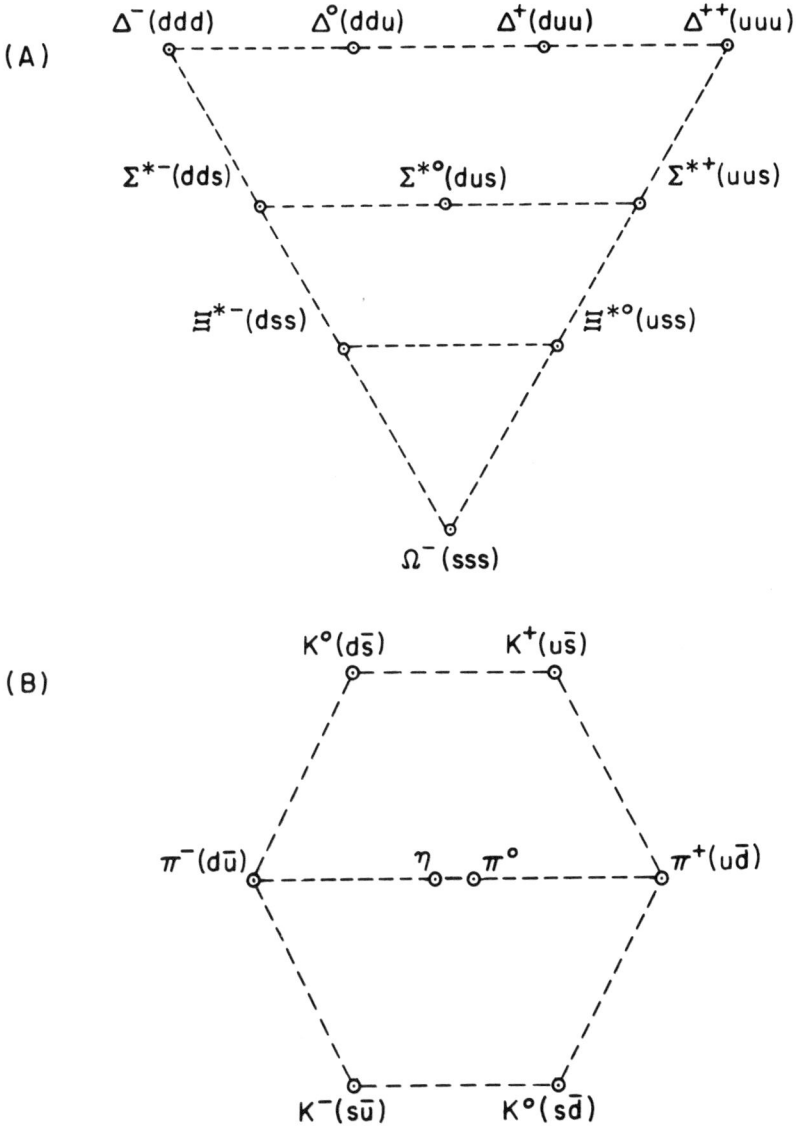

FIGURE 3.1 Hadrons are made out of quarks. (A) shows how the delta, sigma-star, xi-star, and omega family of hadrons are made out of three quarks; (B) shows how the meson family, which contains the pion and kaon, is made of a quark and an antiquark. The positive pion, π^+, and the positive kaon, K^+, have different properties because the π^+ consists of an up quark (u) and a down antiquark ($\bar{d}$), while the K^+ consists of an up quark (u) and a strange antiquark ($\bar{s}$). The η and π^0 are made up of combinations of $u\bar{u}$, $d\bar{d}$, and $s\bar{s}$ quarks.

masses of all the Δs indicate that the *u* and *d* quarks have about the same mass.

However, if we were to change one of the *u* quarks in the Δ^{++} into an *s* quark, again without changing the orbit, we would then have a Σ^{*+} (sigma-star plus), which has a mass about 150 MeV greater than the Δ^{++}. This indicates that the *s* quark is about 150 MeV heavier than the *u* or *d* quarks. If we were to change one of the two *u* quarks in the Σ^{*+} into an *s*, we would get the Ξ^{*0} (xi-star zero), about 150 MeV heavier than the Σ^{*+}. And finally, if we were to change the remaining *u* into an *s*, we would have the Ω^- (omega minus). The Ω^- had not been seen when the quark model was first proposed. Its discovery the following year, with the predicted mass and the predicted charge, gave strong support to the quark picture.

But even then many physicists emphasized that the symmetry did not necessarily imply the actual physical existence of quarks. In particular, the charge of the quarks had to be fractional (2/3 of the standard unit for the *u* quark and $-1/3$ for the *d* and *s* quarks), but no fractionally charged particles had ever been observed. Thus although the hadron classification scheme based on quarks was widely accepted, the actual physical existence of quarks was questioned.

The Discovery of the Charmed Quark

During the years from 1964 through 1973, considerable progress was made, both experimentally and theoretically, in support of the idea of physical quarks. Some of this is described in Chapter 3 in the section on How Quarks Interact. But perhaps the most important and compelling new evidence for quarks began in 1974 with the discovery of a new particle, the J/ψ ("jay-psi"), which was discovered simultaneously at Brookhaven National Laboratory (where it was called the *J*) and at Stanford Linear Accelerator Center (where it was called the ψ).

The J/ψ was unusually heavy (3.1 GeV in mass) and had a very long lifetime, uncharacteristic of strongly interacting particles. Indeed, heavy particles in general tend to be more unstable and therefore to have shorter lifetimes. Thus the J/ψ definitely did not fit into the symmetry scheme that had been so successful in classifying other hadrons.

Physicists hypothesized that it contained a new kind of quark, called *c* or charm, which had in fact been predicted earlier. The J/ψ was believed to be a bound state of a charmed quark and a charmed antiquark. In order for the J/ψ to have such a large mass, the mass of the new quark would also have to be large (about 1.5 GeV). Thus the mass of the J/ψ would be about the mass of a c quark plus the mass of a $\bar{c}$ antiquark. [An antiparticle is often symbolized by drawing a short bar above the symbol for the corresponding particle. Thus $\bar{c}$ (pronounced c bar) is the symbol for the charmed antiquark.]

If this hypothesis were correct, it would mean that a whole family of new charmed particles would exist, consisting of a charmed quark bound together with one or more other kinds of quarks. For example, there would be a $c\bar{u}$ state (called the D^0, with a mass of about 1.8 GeV), a $c\bar{d}$ state (called the D^+, with a similar mass), a $c\bar{s}$ (called the F^+, with a mass of about 2.0 GeV), and a uuc state (a charmed baryon, with a mass of about 2.2 GeV).

All these states, and others, have since been discovered! All have had the masses, charges, decay modes, and other properties predicted from the idea of constituent quarks. The excellent agreement between prediction and experiment has established the validity of the quark picture beyond any reasonable doubt.

Charmonium States

The discovery of the J/ψ was also important in establishing the existence of quarks in a second way, since it was the first of several states, referred to as "charmonium" states, that are composed of a $c\bar{c}$ pair. All these states have masses in the range 3.0 to 3.6 GeV. All are believed to consist of a charmed quark bound together with a charmed antiquark. The heavier ones are excited states in the sense that the two quarks have more energetic orbits. The existence of these distinct but similar particles, each formed by the same constituent quarks but in different energy states, provided an important quantitative confirmation that quarks do indeed exist.

Such a range of different energy states in a two-body system is very familiar to physicists. An analogous two-body system is the hydrogen atom, composed of a single electron orbiting around a single proton. The different energy levels of the excited states of hydrogen account for the discrete lines in the spectrum of light emitted by hydrogen; the spectral lines are produced by photons emitted in a transition from an excited level to a less-excited level, and their energy is equal to the difference in energy levels of the initial and final states. Spectral lines were first observed in 1802, and the spectrum of excited states was first quantitatively explained by the Bohr model of the hydrogen atom in 1913.

A similar set of different energy levels is seen in positronium, which is a bound state of an electron and its antiparticle, the positron. Since

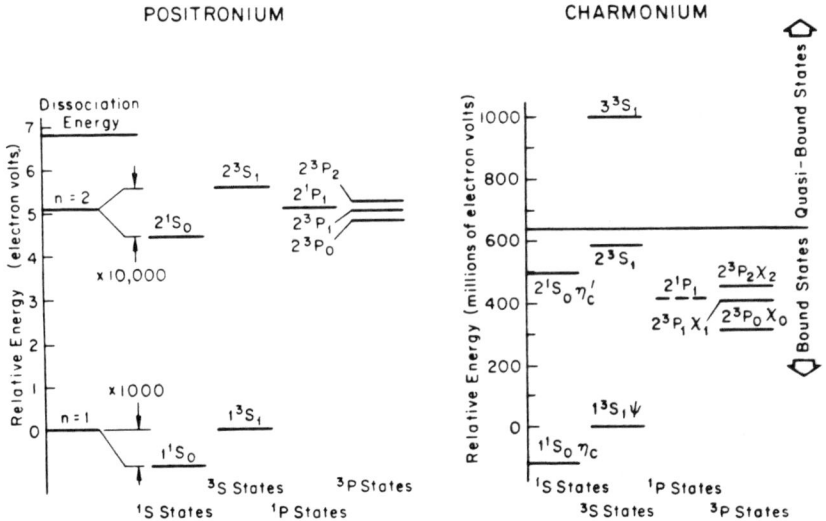

FIGURE 3.2   The spectrum of energy states is similar in positronium and charmonium, but the scale of the energy differences in charmonium is greater by a factor of roughly 100 million. The energy of a state is determined by the principal quantum number n and by the orientation of the particle spins and the orbital angular momentum. The arrangement of the energy levels is similar because both pairs of particles obey the same laws of quantum mechanics. In positronium the various combinations of angular momentum cause only minuscule shifts in energy (shown by expanding the vertical scale), but in charmonium the shifts are much larger. All energies are given with reference to the 1^3S_1 state. At 6.8 electron volts positronium dissociates. At 633 MeV above the energy of the ψ charmonium becomes quasi-bound because it can decay into D^0 and $\bar{D}^0$ mesons.

charmonium states are also bound states of a particle (the c quark) and its antiparticle, they should show a spectrum of energy levels similar to those of positronium. However, since the charmed quark is about 3000 times more massive than the electron, and since the force holding the quarks together in charmonium is the nuclear force (about 100 times stronger than the electric force), one would expect the masses and the mass difference between charmonium states to be much larger than those of the positronium states.

This is exactly what is observed. Seven different charmonium bound states have been found. These states are shown in Figure 3.2(b). The similar states for positronium are shown in Figure 3.2(a). Note that the energy spacing between the charmonium levels is about 100 million times larger than the spacing between the positronium levels. But aside from this expected difference, the close similarity of the structure of the splittings speaks for itself and provides another strong proof of the physical existence of quarks and of the universality of quantum mechanics.

DISCOVERY OF THE THIRD GENERATION OF LEPTONS AND QUARKS

With the discovery of the charmed quark in 1974, the second generation of quarks was completed. At that time, two generations of leptons were also known: the electron and its neutrino, and the muon and its neutrino. It is interesting to go back to 1974 to understand the significance of the two generations and to give a brief history of how the third generation was accidentally discovered in both the lepton and the quark areas. In 1974 there was no explanation of why there was more than one generation of either leptons or quarks, and indeed we still have no explanation of this fact. As discussed in the next chapter, this is one of the outstanding puzzles facing elementary-particle physicists.

The Discovery of the Tau Lepton

The generations puzzle is most easily seen in terms of the charged-lepton situation in 1974. At that time we knew that both the electron and the muon existed, that the muon was about 200 times heavier than the electron, and that both the muon and the electron had the same kind of behavior with respect to the electromagnetic force and the weak force. We also knew that the muon was very different from the electron in the sense that it could not decay into an electron in any simple way. But there was absolutely no theoretical understanding of why both particles existed or of how the mass of the muon was related to the mass of the electron.

Experimenters at the SPEAR electron-positron collider at the Stanford Linear Accelerator Center (SLAC) began to look at the particles being produced in this machine to see if there might be charged leptons other than the electron or muon being created in the collisions. This was purely an experimental search, since there was no theoretical motivation for it. This is an illustration of a theme that we shall return to again and again in this report—that experimenters often explore the unknown without theoretical guidance. And such explorations can be very fruitful, particularly at new accelerator facilities. SPEAR was such a facility in 1974.

In 1975 these experimenters began to accumulate evidence for the existence of the third charged lepton, now called the tau. The tau has

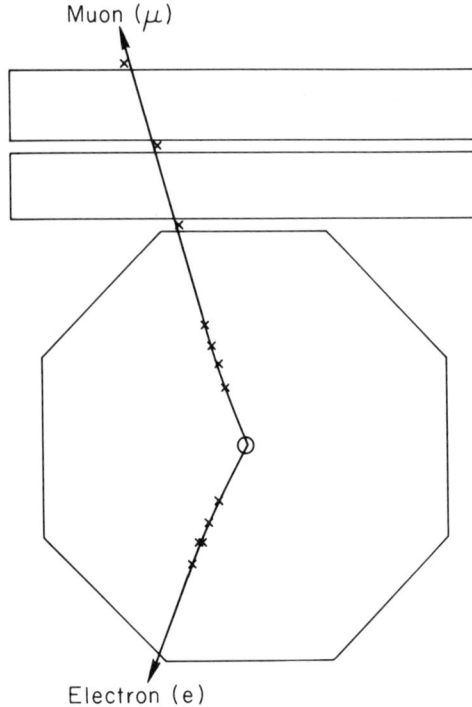

FIGURE 3.3   One of the electron-muon two-prong events that led to the discovery of the tau lepton in 1975. At the time such events were unusual and could not be explained by the production of any of the then known particles.

a mass of a little over 1780 MeV; hence it is about 3500 times heavier than the electron. The discovery was made through the finding of electron-muon two-charged-particle events as shown in Figure 3.3. The tau lepton had too short a lifetime to be detected directly at that time, but in an electron-positron collision a tau-antitau pair can be produced, and this pair can then decay to an electron and a muon, plus unseen neutrinos.

Subsequent studies of the tau lepton at SPEAR and other electron-positron colliders showed that it behaved the same way as the electron and muon with respect to the weak and electromagnetic force and that it did not respond to the strong force.

Further studies of the decay of the tau lepton demonstrated that it had its own unique neutrino associated with it. That is, the neutrino associated with the tau lepton is not the same as the neutrino associated with the electron, nor as the neutrino associated with a muon. Thus two new leptons were actually found, the tau lepton and its associated neutrino.

It is still necessary for us to learn how the tau neutrino interacts, i.e., to see if it interacts in a manner similar to the way in which the electron neutrino and the muon neutrino interact. Such an experiment cannot be carried out in an electron-positron collider, where all other studies of the tau and its neutrino have been done, but rather must make use of a secondary neutrino beam produced at a proton accelerator.

The Discovery of the Bottom Quark

The discovery of the *b* or bottom quark was made at Fermilab in 1977. As in the case of the tau, this was a purely experimental discovery. There was little theoretical guidance in looking for the *b* quark and no indication of what energy might be required to find it. The experiment at Fermilab that found the *b* quark was studying pairs of electrons and pairs of muons produced in the collisions of the primary proton beam of the 400-GeV proton accelerator with a fixed target. The experimenters measured the masses of the pairs of electrons or muons produced, and they plotted the frequency of occurrence of those masses, as shown in the historic curve of Figure 3.4. A peak in that mass frequency plot appears between 9 and 10 GeV.

This peak turned out to be due to the production of a new kind of particle called the upsilon. Each of the upsilon particles consists of a bottom quark bound together with its corresponding antiquark. Hence the mass of the bottom quark is about half of 10 GeV, namely, 5 GeV. This is how the bottom quark was discovered.

Information about the bottom quark can be obtained by studying the upsilon family of particles or by studying mesons that consist of one bottom quark and one of the lighter antiquarks (or vice versa). Such particles are called *B* mesons. Extensive studies of upsilon particles and *B* mesons have been and are being made, particularly at electron-positron colliders. For example, Figure 3.5 shows the spectrum of the upsilon family of particles, obtained at the Cornell Electron Storage Ring (CESR) and DORIS [at the Deutsches Electronen Synchrotron (DESY)] electron-positron colliders.

B mesons are probably also copiously produced in hadron-hadron collisions, either in fixed-target experiments or at particle colliders. At present, the large background of ordinary mesons also produced in hadron-hadron collisions makes the detailed study of *B* mesons difficult when produced in this way. But as particle detectors improve, it should become possible to make detailed studies of *B* mesons at proton

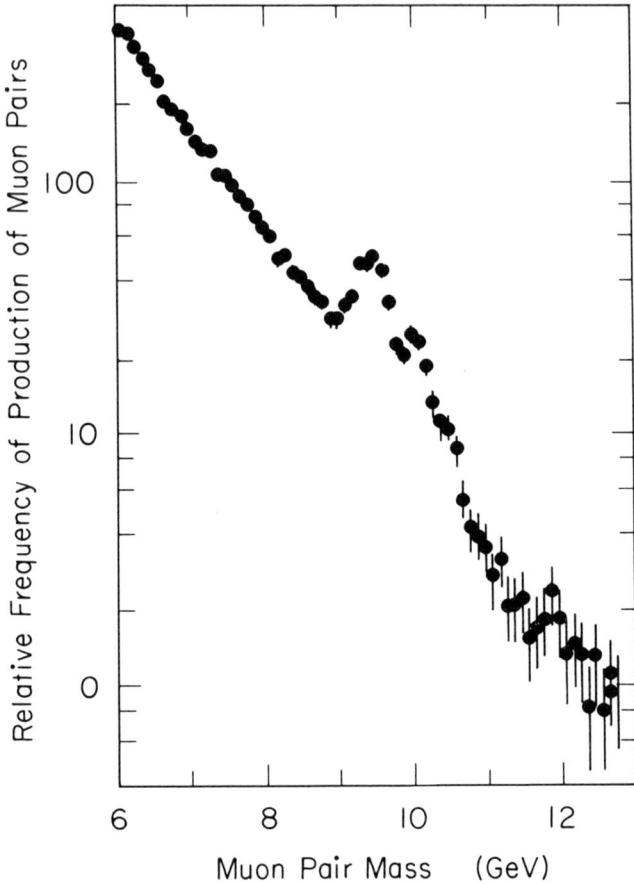

FIGURE 3.4 The upsilon was discovered in 1977 by studying the production of muon pairs or electron pairs in proton collisions. Here the relative frequency of production of muon pairs is shown to decrease as the muon pair mass increases. The bump in the curve at 9-10 GeV is due to the upsilon.

accelerators as well as those currently done at electron-positron colliders.

The Third Generation

As shown in Figure 3.6, we can now see how the third generation of leptons and half of the third generation of quarks was added to our basic system of elementary particles. Most physicists believe that there is a second member of the third generation of quarks, which is called the *t* or top quark. The expectation for the existence of the top quark comes from two sources: first is our belief that nature is simple, so that in each generation quarks like leptons should come in pairs; second, measurements of the *b* quark lifetime give an indirect indication that there should be a top quark associated with the bottom quark. As this report was being completed in 1984, initial direct evidence was reported for the existence of the top quark.

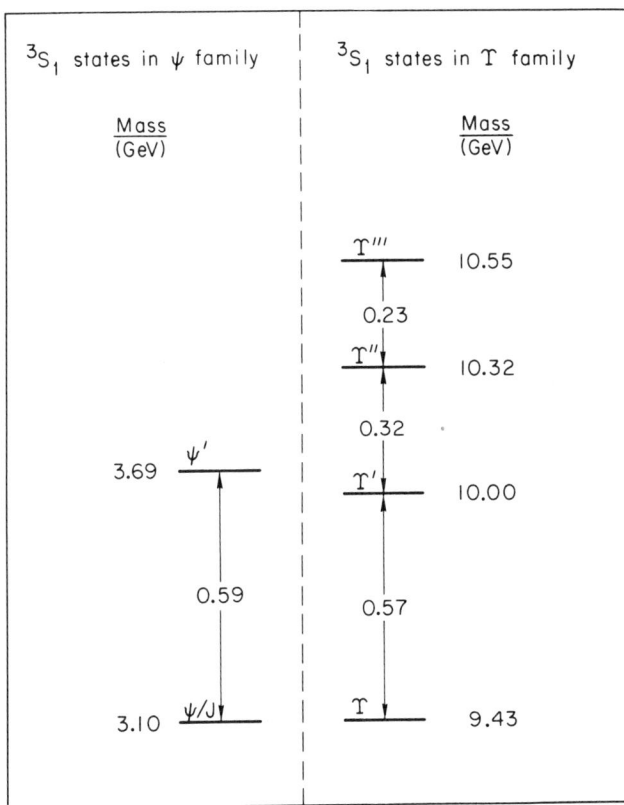

FIGURE 3.5   The triplet S states ($^3S_1$) of the upsilon (Υ) family are shown on the right. Each of these states consists of a b quark bound to a $\bar{b}$ quark. For comparison the two $^3S_1$ states of the ψ family are shown on the left. Although the masses are very different, the level separations are nearly equal.

Generation	Particle	Charge	Mass

| I | electron (e) | −1 | 0.51 MeV |
| | electron neutrino (ν_e) | 0 | less than 50 eV |

| 2 | muon (μ) | −1 | 106 MeV = 0.106 GeV |
| | muon neutrino (ν_μ) | 0 | less than 0.5 MeV |

| 3 | tau (τ) | −1 | 1784 MeV = 1.784 GeV |
| | tau neutrino*(ν_τ) | 0 | less than 160 MeV = 0.160 GeV |

*indirect evidence

Generation	Particle	Charge	Mass

| I | up (u) | +2/3 | about 300 MeV = 0.3 GeV |
| | down (d) | −1/3 | about 300 MeV = 0.3 GeV |

| 2 | charm (c) | +2/3 | about 1500 MeV = 1.5 GeV |
| | strange (s) | −1/3 | about 500 MeV = 0.5 GeV |

| 3 | | | |
| | bottom (b) | −1/3 | about 5,000 MeV = 5.0 GeV |

FIGURE 3.6   Our present knowledge of the lepton and quark families of particles.

Although nature does seem to be simple, that does not mean that we understand it. Just as in 1974 we did not know why there were two generations of leptons and quarks, so in 1985 we do not know why there are three generations of leptons and quarks. What has been gained, of course, is the experimental demonstration that there can be more than two generations of leptons and quarks. Indeed, there may be more than the present three generations. Some theoretical arguments and some deductions from astrophysical considerations can be interpreted to mean that there are not more than four generations of leptons and quarks. But physics is, in the end, an experimental science, and the search for more than four generations of leptons and quarks will be carried on by experimenters. There is probably nothing more challenging to a scientist than to be told that, theoretically, something cannot exist.

HOW QUARKS INTERACT

Hadron Interactions

A large body of systematic knowledge has been developed as a result of many experiments on the interactions of hadrons with each other at different energies. Such interactions include elastic scattering, where one hadron simply bounces off the other, with neither hadron being changed, and inelastic scattering, where more hadrons are created in the collision. The basic quantitative understanding of this vast body of data has so far been limited to some areas where the new theory of quantum chromodynamics can be applied, as discussed in Chapter 3 in the section on Strong Interactions among Quarks. However, the systematics and qualitative behavior of hadron collisions have been a valuable guide in the understanding of the quark structure of hadrons.

The understanding of the quark structure of hadrons proceeds most easily from considering not the collision of two hadrons but rather the collision of a lepton with a hadron. In this case we consider the collision of a simple particle, the lepton, with a complicated particle, the hadron. In practice the hadron is either a proton or a neutron.

Lepton-Proton Scattering Experiments

Isolated free quarks apparently do not exist in nature; they always seem to be bound within hadrons. Yet it is nevertheless possible to see an individual quark inside a hadron. This was first done at SLAC in 1969, long before the c and b quarks were discovered, by scattering high-energy electrons off protons.

This scattering process occurs through the exchange of a single photon between the electron and proton, as shown in Figure 3.7(a). This interaction will generally produce a multiparticle shower of hadrons if the electron has high enough energy. This shower is extraordinarily complex and difficult to describe mathematically.

Now let us picture what must be happening in the interaction if the proton is composed of quarks. Since the quarks carry all the charge in the proton, the photon must interact with one of these quarks. The fundamental interaction between the electron and the struck quark is therefore a simple electromagnetic scatter, as shown in Figure 3.7(b). If the spectator quarks are disregarded, the interaction in Figure 3.7(b) is identical to that in Figure 3.7(a) and is of the type that can be calculated using the well-established rules of quantum electrodynamics.

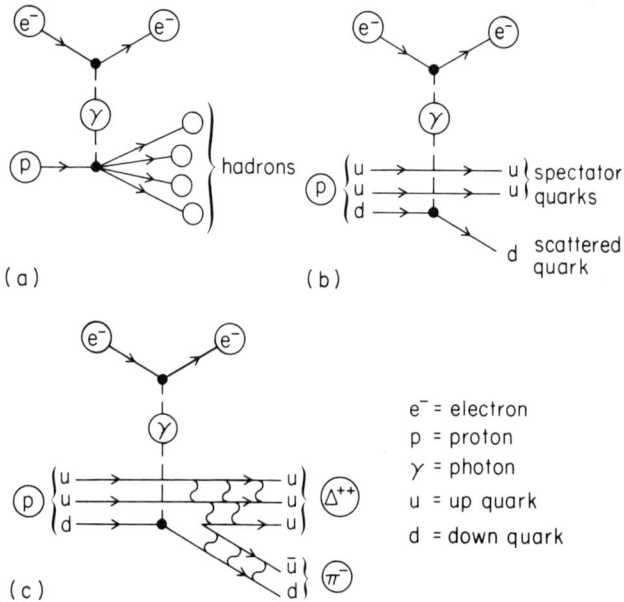

FIGURE 3.7   (a) In the inelastic scattering of an electron and a proton, the electron itself does not interact with the proton. A photon, indicated by the dashed lines, is emitted by the electron and interacts with the proton producing hadrons. (b) A more detailed description of how an electron inelastically scatters from a proton. The photon emitted by an electron interacts with one of the quarks in the proton. The other two quarks are hardly affected and are called spectator quarks. (c) An example of what can finally happen to the quarks in an inelastic scattering. In this example an additional u quark and $\bar{u}$ antiquark pair are produced. The three u quarks unite to form a delta hadron (Δ^{++}), and the other two quarks form a pion (π^-). To the best of our knowledge free quarks never escape from an inelastic scattering but always unite somehow to form hadrons. The wiggly lines are gluons that carry the strong force between the quarks.

When the struck quark is knocked away from the two spectator quarks, hadrons are produced by the strong interaction between the scattered quark and the spectator quarks. An example is shown in Figure 3.7(c), where Δ^{++} and π^- hadrons are produced. But the electron never sees these interactions between the quarks in the final hadronic system. As far as the electron is concerned, the interaction is a simple scattering process with a single apparently free quark.

It is not difficult to prove mathematically that if both the energy and the deflection angle of the scattered electron are measured, the momentum of the struck quark can be calculated from these. Thus it is possible to determine the momentum distribution of quarks within the proton by measuring the angular distribution of final-state electrons of a given energy.

An important consequence of this picture is that the angular distributions of scattered electrons measured for two different interaction energies are closely related, since they must originate from the same quark distribution. This relation, known as scaling, was experimentally observed in the SLAC electron-proton scattering experiments and strongly supported the idea of physical quarks.

The interaction of a neutrino with a proton occurs through the weak force rather than through the electromagnetic force, but as shown in Figure 3.8 it is otherwise a similar process. Here the incident neutrino turns into a μ^-, emitting a W^+ in the process; the W^+ is one of the carriers of the weak force. The W^+ then hits a d quark in the nucleon, which changes into a u quark when it absorbs the W^+. The weak interaction thus changes the type of the interacting quark. But the momentum distribution of quarks within the proton is revealed by the weak interaction in the same way that it is revealed by the electromagnetic interaction in Figure 3.7(b).

The momentum distribution of quarks in a high-energy proton is usually given as the probability of finding a quark that carries a certain

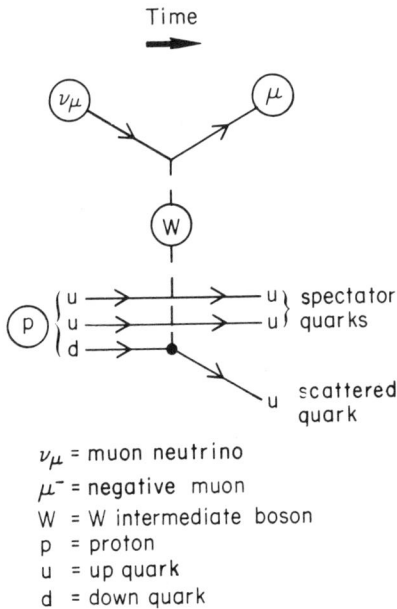

ν_μ = muon neutrino
μ^- = negative muon
W = W intermediate boson
p = proton
u = up quark
d = down quark

FIGURE 3.8 The inelastic scattering of a neutrino on a proton is analogous to the inelastic scattering of an electron on a proton (Figure 3.7). But here the neutrino emits a W particle that interacts with one of the quarks. In addition, when the neutrino emits the W it changes into a muon.

fraction of the proton's momentum. These momentum distributions have been measured in various kinds of experiments, using muon and neutrino beams as well as electron beams. Such experiments have demonstrated that, in addition to the three valence quarks that we expect to see, there is a neutral sea of gluons (which carry the force binding the quarks together) mixed with a sea of low-energy virtual quark-antiquark pairs that are produced by the gluons. Each of these three components—the valence quarks, the gluons, and the virtual quark-antiquark pairs—carries part of the proton's momentum.

By comparing experiments that use different kinds of incident beams (muons, electrons, neutrinos, and antineutrinos) on different kinds of targets (hydrogen, deuterium, and iron, for example) it is possible to do the following things:

1. Measure the distributions of different individual kinds of quarks within the proton.

2. Count the total number of valence quarks within the nucleon (expected to be 3).

3. Measure the mean-square charge of the quarks in the nucleon.

All these measurements agree with the values expected from the quark model.

Evidence of the underlying quark structure of hadrons can be found in many different kinds of experiments. An interesting example is the production of $\mu^+\mu^-$ pairs in hadronic collisions. Since the μ^+ and μ^- are leptons, they cannot interact through the strong force and thus are usually not produced in hadronic collisions. But occasionally a quark in one of the hadrons will electromagnetically annihilate with an antiquark in the other hadron, producing a massive photon that decays into $\mu^+\mu^-$. The process is similar to that in which an electron and a positron annihilate and produce a $\mu^+\mu^-$ pair, except that the colliding particles are now quarks instead of leptons. The rate of $\mu^+\mu^-$ production, as well as the distributions of the muon pairs as a function of energy and production angle, generally agree with the predictions from quark-antiquark annihilation.

Hadron Jets

Perhaps the most striking way of seeing evidence of individual quarks is in the production of hadrons through electron-positron annihilation. Such interactions are observed in colliding-beam experiments at the PEP storage ring at SLAC and at the PETRA storage ring at DESY. The production process, Figure 3.9(a), occurs in two steps:

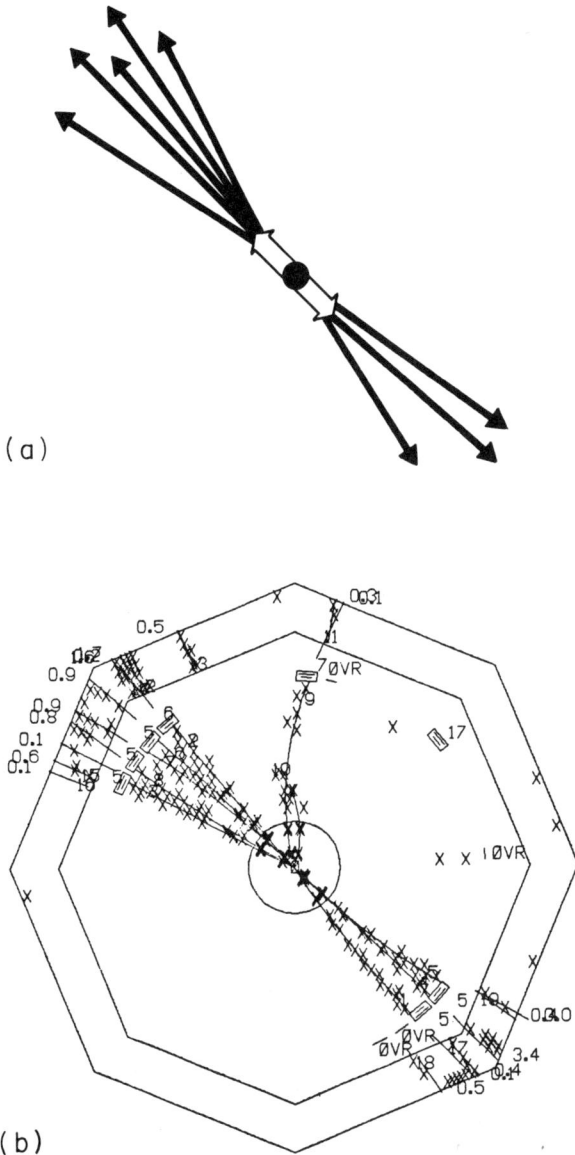

FIGURE 3.9   In (a) the central black dot represents the point where the electron and positron annihilated. The open arrows represent a quark and an antiquark produced in that annihilation. The quark and antiquark begin to move in opposite directions. But as they separate they each change into a shower or jet of hadrons. This picture has been drawn to show the jets of hadrons moving in the same directions as in an actual event (b) obtained at PEP.

(1) The colliding e^+ and e^- electromagnetically annihilate to produce a quark-antiquark pair, just as $\mu^+\mu^-$ pairs can be produced in the identical process. (2) As the quark and antiquark separate, the strong force between them builds up energy, which is transformed into more quark-antiquark pairs. These quarks and antiquarks then coalesce to form hadrons, as described earlier. The striking feature of these events is that, because of the string nature of the strong force, the new hadrons tend to be produced along the line joining the two originally separating quarks. This results in the hadrons appearing in two back-to-back jets of particles, which follow the directions of the original quark and antiquark. As the energy of the interaction is increased, the jets become more collimated and contain more particles. Figure 3.9(b) shows a typical e^+e^- interaction observed in a high-energy experiment at PEP.

A similar effect occurs in proton-proton or in proton-antiproton interactions at very high energies. Particles with high momentum perpendicular to the beam direction are produced predominantly by the collision of two quarks (or a quark and an antiquark, or a gluon and a quark, or two gluons) giving two jets in the final state. In the case of proton-antiproton collisions, the two spectator quarks in the proton and antiproton also form jets. This gives a total of four jets, two along the direction of the original colliding particles and two more in the directions of the scattered quarks.

The details of the process through which the scattered quarks form hadrons, called hadronization, cannot be exactly calculated yet because of its great complexity. However, phenomenological and approximate methods have been used to compare hadronization in different kinds of production processes, and these have been successful in relating production rates of many different kinds of particles in high-energy interactions between pions, kaons, and protons.

The concept of quarks, and the understanding of how hadrons are composed of quarks and of how quarks interact, has vastly furthered our understanding of the nature of matter. In addition, the quark substructure of matter is revealed in all the different kinds of interactions—electromagnetic, weak, and strong. Thus the idea of quarks has led to a great simplification in the way in which we understand the interactions of hadrons. We are now able to recognize and to study the fundamental interaction (involving quarks) within the apparent interaction (involving hadrons). This makes it possible to focus our attention on these fundamental processes, and thus to measure and understand the characteristics of the fundamental forces at the most basic level.

UNIFICATION OF THE WEAK AND ELECTROMAGNETIC INTERACTIONS

The force of electromagnetism shapes the world around us. The structure of matter, the chemistry of life, and the propagation of light all may be traced to the basic laws of electrodynamics. Electricity and magnetism, encountered in everyday experience as the spark of a static discharge and the gentle swing of a compass needle, would seem to be quite distinct phenomena. But a long line of brilliant nineteenth-century experiments showed them to be two different facets of the same underlying interaction. This set the stage for Maxwell's 1862 unification of electromagnetism in simple equations that embodied all the understanding of the nature of light, indicated the possibility of radio communication, and was the starting point for the development of quantum electrodynamics.

Quantum electrodynamics (QED) is the most successful of physical theories. It has achieved predictions of enormous accuracy, such as that of the anomalous magnetic moment of the electron, for which theory and measurement agree to at least seven decimal places. Such detailed predictions have stimulated, and been stimulated by, experiments of remarkable inventiveness and precision. Moreover, the predictions of QED have been verified over an extraordinary range of distances, from less than 10^{-18} m (a billionth of a billionth of a meter) to more than 10^8 m (100 million meters).

It is therefore natural that QED should serve as a model for other theories. The earliest attempt at a description of the weak interactions, due to Fermi in 1933, was constructed by direct analogy with QED. Much subsequent work has involved extending this analogy and determining its limits of applicability. By 1957, when it was established that the weak interactions were intrinsically left-handed, and not mirror-symmetric like electromagnetism, an extremely successful operational description of radioactivity and related weak-interaction processes had been achieved.

A second aspect of theoretical work has been the idea of a synthesis, following the example of electromagnetism. Having profited from the idea that the weak and electromagnetic interactions are at least analogous, one is prompted to ask whether they might actually be related. In relativistic quantum theories, interactions are mediated or carried by force particles. The carrier of the electromagnetic interaction, the photon, was postulated in 1905 by Einstein. Its existence was confirmed in the 1920s by experiments that showed that light scattered like massless particles from electrons. It was appealing to hypothesize

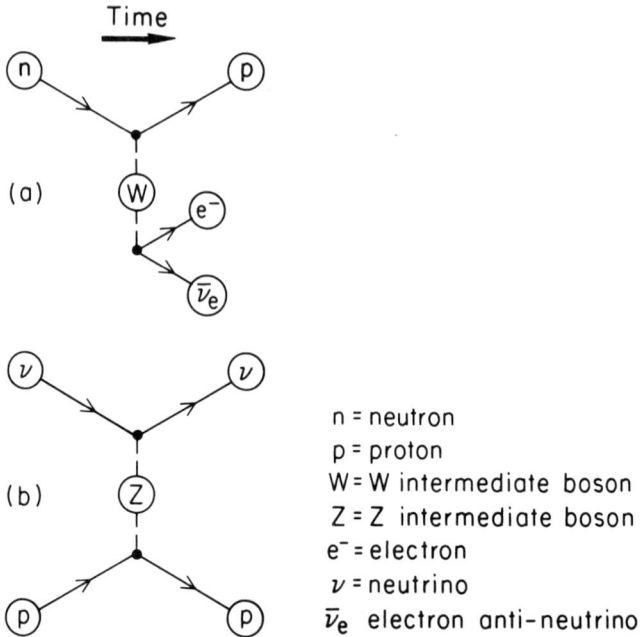

FIGURE 3.10   Two examples of processes that take place through the weak interaction. In (a) a neutron decays into a proton. It does this by emitting a *W* particle; the *W* itself is unstable and decays into an electron and a neutrino. The *W* carries or mediates the weak force, hence it is called an intermediate boson. In (b) a neutrino scatters off a proton by emitting a *Z* particle. This process is analogous to the scattering of an electron on a muon (Figure 2.8). The *Z* is also an intermediate boson since it carries the weak force.

that the weak interaction is carried by a so-called intermediate boson, denoted *W* for weak. This weak boson must be electrically charged in order to mediate nuclear radioactive decays such as the disintegration of a neutron into a proton, an electron, and an antineutrino, as shown in Figure 3.10(a). It was apparent from early investigations of natural radioactivity that the conjectured intermediate boson must be extremely massive in order to explain the long lifetimes that were observed. The idea that the weak and electromagnetic interactions—so different in apparent strength—have a common origin provides an estimate of the *W*'s mass of approximately 100 times the proton mass.

To advance from these general notions of analogy and synthesis to a viable theory of the weak and electromagnetic interactions has required a half century of experimental discoveries and precision measurements and of theoretical insights and inventions. Like QED itself, the unified theory is a gauge theory derived from a symmetry principle. In this case, the symmetry is a family pattern among quarks and leptons that was suggested by experiments. A self-consistent theory could not be based on the known force particles (the photon and the conjectured *W*) alone but required in addition an electrically neutral weak force particle Z^0 and an auxiliary object known as the Higgs particle. The latter plays a key role in hiding the electroweak symmetry. This is required to account for the varied masses of the quarks and leptons.

Just as the *W* particles mediate charge-changing transitions such as neutron decay, the Z^0 must mediate a new class of neutral-current weak interactions such as neutrino-proton elastic scattering, shown in Figure 3.10(b). At the time that the theory was formulated, there was no experimental evidence for neutral-current interactions. The discovery of a few characteristic events in the Gargamelle bubble chamber at CERN in 1973, quickly supported by evidence from Fermilab, Brookhaven, and Argonne, marked the beginning of an intensive study of this new phenomenon. An example of a neutral current event is given in Figure 3.11.

A decade of experimentation with high-energy neutrino beams, together with important results from electron scattering at SLAC and from electron-positron annihilations at PETRA and PEP, has shown the neutral-current interaction to behave as expected in electroweak theory. The experiments using electrons involved both the electromagnetic force and the weak force. These two forces, occurring in the same experiment, interfere with each other. The detection of these interference effects was one of the first confirmations of the correctness of the unified theory.

It remained to observe the intermediate bosons as real (though ephemeral) particles, rather than merely seeing the interactions attributed to their existence. In the model, the properties of the intermediate boson, such as their masses, depend on a single parameter that has been determined from neutral-current experiments. On this basis, we expected the mass of the charged intermediate boson *W* to be about 83 GeV/c^2 and the mass of the neutral intermediate boson Z^0 to be about 95 GeV/c^2. Both charged and neutral bosons should disintegrate less than a trillionth of a trillionth of a second after formation. Such prodigious masses are attainable only in colliding-beam machines, specifically at present in the proton-antiproton collider operating at CERN.

Collisions of protons and antiprotons result in interactions among their constituent quarks, antiquarks, and gluons. Because the way in

68

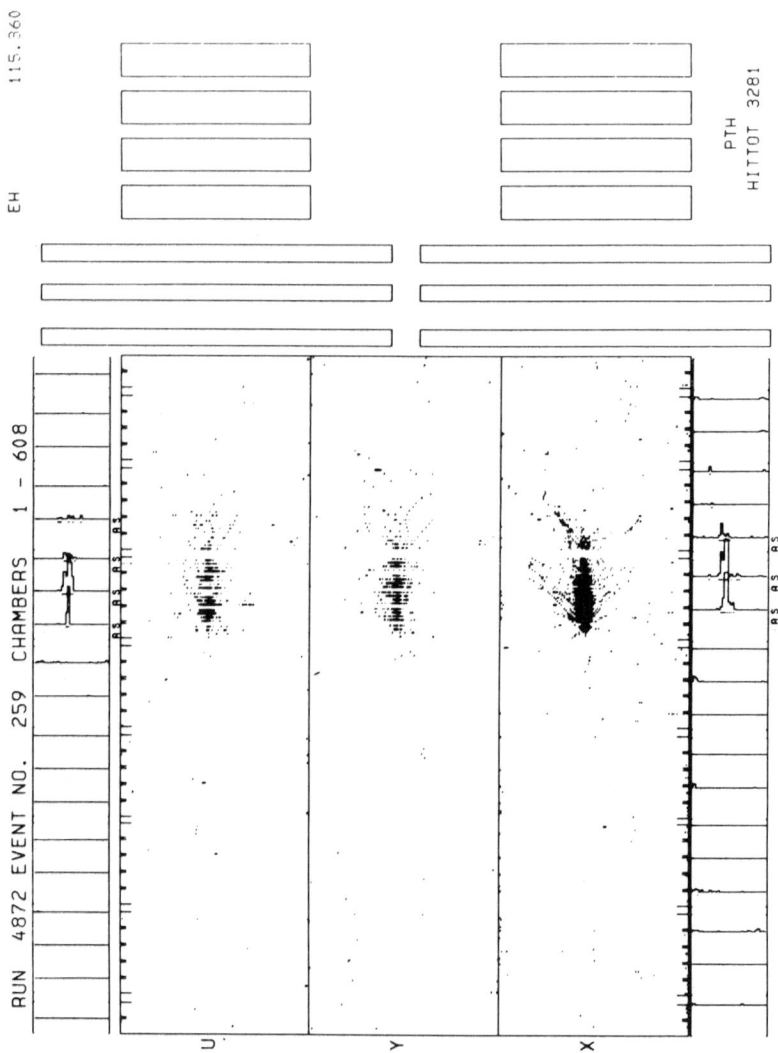

FIGURE 3.11 A neutral current event recorded at Fermilab. The incident neutrino enters from the left; the absence of a penetrating particle in the detector indicates that there is no muon in the final state. Therefore the neutrino is not interacting through the charged current.

Time

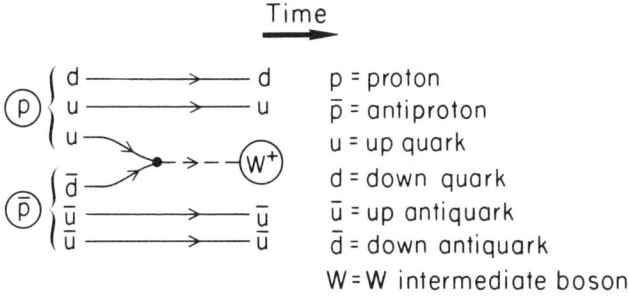

```
        ⎧ d ————————→—— d          p = proton
   ⓟ   ⎨ u ————————→—— u          p̄ = antiproton
        ⎩ u ——⟍                    u = up quark
              ⟍——→—●—>--(W⁺)
        ⎧ d̄ ——⟋                    d = down quark
   ⓟ̄   ⎨ ū ————————→—— ū          ū = up antiquark
        ⎩ ū ————————→—— ū          d̄ = down antiquark
                                    W = W intermediate boson
```

FIGURE 3.12 The dominant process for production of the W intermediate boson in proton-antiproton collisions. A u quark from the proton (p) and a $\bar{d}$ antiquark from the antiproton ($\bar{p}$) unite to form the W^+. The spectator quarks form into new hadrons, which are not shown.

which quarks and antiquarks should combine to form intermediate bosons is known and the motion of quarks within the proton has been extensively studied, we can calculate that one intermediate boson will be produced in about 5 million proton-antiproton collisions. The dominant production mechanism is shown in Figure 3.12. To extract the intermediate bosons from the background of ordinary events requires an elaborate detector that can recognize and measure the characteristic decay products amid the debris of a violent collision. The most characteristic signal for W decay is an energetic electron emitted transverse to the direction of the colliding beams and an undetected neutrino with equal and opposite transverse momentum. In the case of the Z^0, a back-to-back electron and positron (antielectron) provide an unmistakable pattern. Both of these particles have in fact recently been observed in the CERN collider experiments. On initial evidence, they have the masses and other properties predicted by the electroweak theory.

This successful search is the culmination of 50 years of speculation on intermediate bosons. The results represent impressive triumphs of accelerator art, experimental technique, and theoretical reasoning. They indicate that the basic electroweak symmetry scheme is correct. More detailed studies of the intermediate bosons and their decay products will be high on the agenda for future experiments at the CERN collider and the Fermilab Tevatron. Electron-positron colliders to serve as Z^0 factories with an annual output of a million Zs or more have been initiated at SLAC (Stanford Linear Collider) and at CERN (LEP). The quest for the Higgs boson or a symmetry-breaking mechanism is the most pressing open issue in electroweak physics.

STRONG INTERACTION AMONG QUARKS

We have seen already how the idea that the strongly interacting particles are built up of quarks brought new order to hadron spectroscopy and suggested new relations among mesons and baryons. But this constituent description also brought with it a number of puzzles. These seemed at first to indicate that the quark model was nothing more than a convenient mnemonic recipe. In pursuing and resolving these puzzles, physicists have found a dynamical basis for the quark model that promises to give a complete description of the strong interactions.

An obvious question concerns the rules by which the hadrons are built up out of quarks. Mesons are composed of one quark and one antiquark, while baryons are made of three quarks. What prevents two-quark or four-quark combinations? Within this innocent question lurks a serious problem of principle. The Pauli exclusion principle of quantum mechanics is the basis for our understanding of the periodic table of the elements. It restricts the configurations of electrons within atoms and of protons and neutrons within nuclei. We should expect it to be a reliable guide to the spectrum of hadrons as well. But according to the Pauli principle, the observed baryons such as Δ^{++} (*uuu*) and Ω^- (*sss*), which would be composed of three identical quarks in the same state, cannot exist.

To comply with the Pauli principle, it is necessary to make the three otherwise identical quarks distinguishable by supposing that every type of quark exists in three varieties, fancifully labeled by the colors red, green, and blue. Then each baryon can be constructed as a colorless (or white) state of a red quark, a green quark, and a blue quark. Similarly, a meson will be a colorless quark-antiquark combination. The rule for constructing hadrons may then be rephrased as the statement that only colorless states can be isolated.

A second issue is raised by the fact that free quarks have not been observed. This suggests that the interaction between quarks must be extraordinarily strong, and perhaps permanently confining. That free quarks are not seen is of course consistent with the idea that colored states cannot exist in isolation. On the other hand, the quark model description of violent collisions rests on the assumption that quarks within hadrons may be regarded as essentially free.

This paradoxical state of affairs may be visualized as follows. We may think of a hadron as a bubble within which the constituent quarks are imprisoned. The quarks move freely within the bubble but cannot escape from it. This picturesque representation yields an operational understanding of many aspects of hadron structure and interactions,

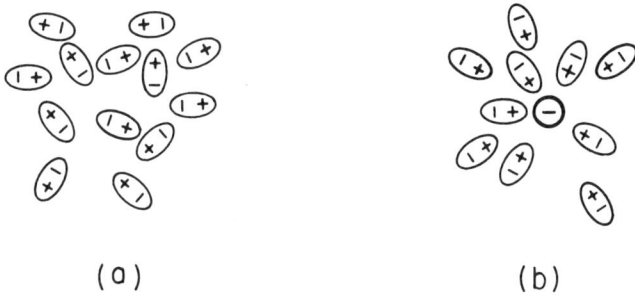

(a)   (b)

FIGURE 3.13   Electrically polarized molecules weaken the effect of an electric charge. In (a) the molecules point in random directions. In (b) a negative charge is present, and the positive ends of the molecules point toward this charge and partially cancel it. Outside of this area the electric charge will appear weaker.

but it falls far short of a dynamical explanation for the puzzling behavior of quarks. We still do not understand completely why quarks apparently interact only weakly when they are close together and yet cannot be pulled apart. To see why this is surprising, and to learn how it might come about, it is helpful to consider the interactions of electrically charged objects.

We customarily speak of the electric charge carried by an object as a fixed and definite quantity, as indeed it is. However, if a charge is placed in surroundings in which other charges are free to move about, the effect of the charge may be modified. An example is a medium composed of many molecules, each of which has a positively charged end and a negatively charged end. In the absence of an intruding charged particle, the molecules are oriented randomly [Figure 3.13(a)], and the medium is electrically neutral not only in the large, but locally as well, down to the submolecular scale. Placing an electron in the medium polarizes the molecules [Figure 3.13(b)]: the negatively charged ends of the molecules are repelled by the electron, while the positively charged ends are attracted to it. The effect of this orientation of the molecules is that at finite distances from the electron its influence is screened, or reduced, by the opposite charges it has attracted. Only when we inspect the electron at very close range—smaller than molecular size—is the full magnitude of the electron's charge apparent. We may say that the effective charge is larger at short distances than at long distances.

We normally think of the vacuum, or empty space, as the essence of nothingness. However, in quantum theory the vacuum is a complicated and seething medium in which virtual pairs of charged particles, most

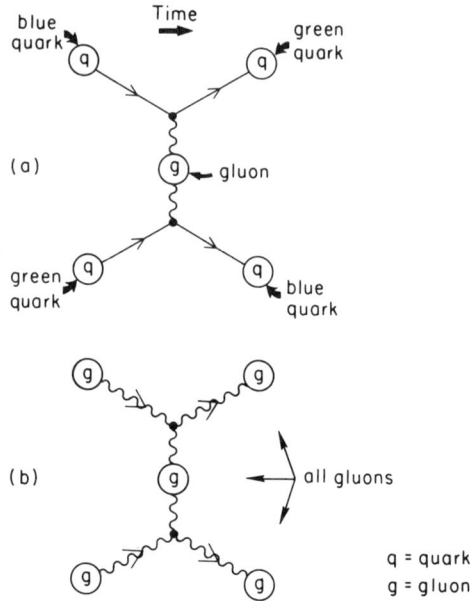

FIGURE 3.14   How quarks and gluons interact. In (a) quarks change their color through the emission and absorption of a gluon. In (b) gluons interact with each other.

importantly electrons and positrons, have a fleeting existence. These ephemeral vacuum fluctuations are polarizable in the same way as the molecules of our example. Consequently in QED it is also expected that the effective electric charge should increase at short distances, and indeed the consequences of this variation are observed in atomic spectra. The behavior of the effective charge in QED is opposite to that required in the realm of the strong interactions, where the interaction between quarks must diminish in strength at short distances.

An explanation for the contrary behavior of the strong force emerged unexpectedly from the ideas that had proved so fruitful for the electroweak interactions: the strategy of gauge theories. Since color is an attribute of quarks but not of leptons, it can be considered as a strong-interaction charge. When the color symmetry among red, blue, and green quarks is taken as the basis for a gauge theory, the resulting interactions among quarks are mediated by force particles called gluons. There are eight gluons corresponding to the distinct color-anticolor combinations. (The white combination corresponding to an equal mixture of red-antired, blue-antiblue, and green-antigreen is not included.) In quantum chromodynamics, or QCD as the theory is called, quarks may interact as shown in Figure 3.14(a), where a blue quark and a green quark exchange color. Because the gluons carry color, they can interact among themselves as well, as shown in Figure 3.14(b). The photons of QED, being electrically neutral, have no such self-interactions.

The fact that the gluons are (color) charged is responsible for the crucial difference between the behavior of the effective charge in QED and QCD. In the strong-interaction theory there are two competing effects: a screening brought about by the color charges in the fluctuating vacuum and a camouflage effect that is not present in QED. The screening, or vacuum polarization, may be understood just as in electrodynamics. This time we think of the vacuum as a collection of randomly oriented three-cornered objects, as shown in Figure 3.15(a). By placing a green quark in the vacuum, we orient the triangles [Figure 3.15(b)] and screen the color charge.

The behavior of the strong interaction charge is the result of a competition between these two opposing effects. In QCD, the outcome is that the effective color charge does have the properties necessary to reconcile the simple quark model and quark confinement. The steady weakening of the charge at short distances is known as asymptotic freedom because quarks become effectively free at very small separations.

In the regime of short distances probed in violent high-energy collisions, the strong interactions are sufficiently feeble that reaction rates may be calculated using the diagrammatic methods developed for QED. In some measure these calculations reproduce the simple quark model results as first approximations. This is the case, for example, in electron-positron annihilations into hadrons. The quark-antiquark production rate, represented by the diagram in Figure 3.16(a), correctly anticipates both the structure of the dominant two-jet events and the

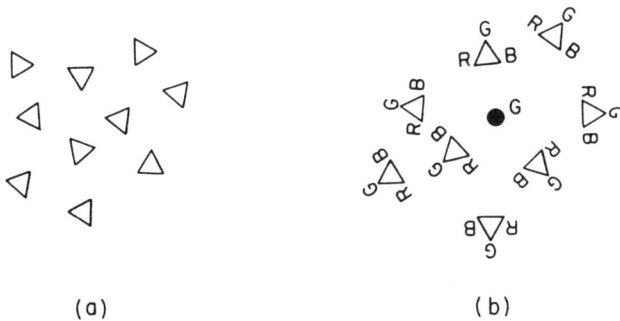

(a)   (b)

FIGURE 3.15 An illustration of how the force exerted by a quark owing to its color charge can be weakened by vacuum effects.

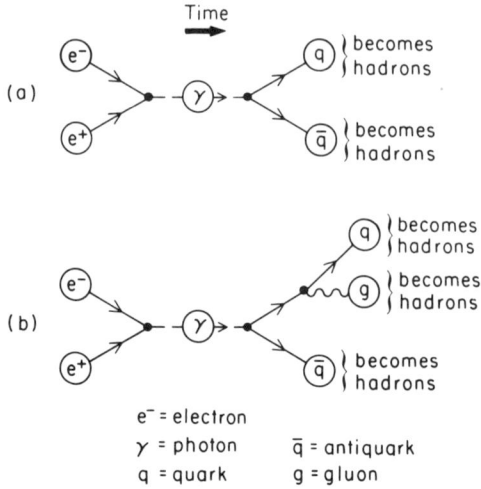

FIGURE 3.16   An example of how interactions of gluons with quarks lead to more complicated processes. In (a) an electron-positron pair annihilate to produce a virtual photon, which in turn produces a quark-antiquark pair. In (b) one of those quarks also emits a gluon.

approximate rate of hadron production. The strong-interaction corrections to this process include the diagram shown in Figure 3.16(b), in which a gluon is radiated by one of the outgoing quarks. Like the quarks, the gluon materializes as a jet of hadrons. The resulting three-jet events are commonplace in the electron-positron annihilations studied at the PEP and PETRA storage rings.

The highest energies yet attained in collisions of the fundamental constituents are those reached in proton-antiproton interactions at the CERN *Sp̄pS* machine. Already collisions among quarks and gluons have been recorded at energies approaching 300 GeV. The hard scatterings of these particles lead to striking jets of hadrons at large angles to the direction defined by the incident proton and antiproton beams. Events of this kind are observed at approximately the frequency suggested by QCD.

While diagrammatic methods are of great value in the study of strong interactions, several considerations prevent the resulting calculations from being as precise as those long familiar in the electroweak domain. The first is that at the energies currently accessible (or, in other words, at the distances currently probed), the strong interaction is still considerably stronger than electromagnetism.

A more serious sticking point is that the diagrammatic methods describe reactions that involve isolated quarks and gluons, whereas in the laboratory quarks and gluons are found only within hadrons. We have not yet succeeded in solving the theory in the regime of potent strong interactions characteristic of hadron structure. In some cases this problem may be circumvented by using QCD only to predict how observables change from one energy to the next and not the value they take at one particular energy. This is the case, for example, in deeply inelastic electron-proton scattering, for which the reaction rate depends in an essential way on the internal structure of the proton. This prediction and observation of gradual but systematic change is one of the notable successes of the theory.

To deal with the existence and properties of the hadrons themselves it is necessary to devise a new computational approach that does not break down when the interaction becomes strong. The most promising approach has been the crystal-lattice formulation of the theory, in which space-time is accorded a discrete, rather than continuous, structure. By considering the values of the color field only on individual lattice sites, one is able to make use of many of the techniques developed in statistical physics for the study of spin systems such as magnetic substances.

One of the most valuable methods has been the use of computer simulations in which different gluon configurations are explored by random sampling (Monte Carlo) techniques. This program makes extremely heavy demands on computer time and has spurred the development and implementation of new computer architectures. Already calculations of this sort have yielded suggestive evidence that quarks and gluons are indeed permanently confined in QCD. Work is continuing actively, with the eventual goal of computing the spectrum and properties of hadrons *ab initio*.

Attempts to understand confinement and the nature of the QCD vacuum have led to the prediction of new phenomena. It seems likely that when hadronic matter is compressed to very great densities and heated to extremely high temperatures hadrons will lose their individual identities. When the hadronic bubbles of our earlier image overlap and merge, quarks and gluons may be free to migrate over great distances. A similar phenomenon occurs when atoms are squashed together in stars. The resulting new state of matter, called quark-gluon plasma, may exist in the cores of collapsing supernovas and neutron stars. The possibility of creating QCD plasma in the laboratory in collisions of energetic heavy ions is under active study.

UNIFIED THEORIES

We have seen in this chapter that developments in elementary-particle physics during the past decade have brought us to a new level of understanding of fundamental physical laws. The establishment—tentative though it be—of the electroweak theory and QCD has brought us a coherent point of view and a single language appropriate for the description of all subnuclear phenomena. The new maturity of elementary-particle physics has made more fruitful the interaction with other areas of physics and promises new insights into the origin of our world.

With QCD and the electroweak theory in hand, what remains to be understood? If both theories are correct, can they also be complete? There are, in fact, many observations that are explained only in part, if at all, by the separate theories of the strong and electroweak interactions. Many of these seem to invite a further unification of the strong, weak, and electromagnetic interactions. This has important consequences not only for our worldview but also for experimental initiatives. Let us examine a few of the patterns unexplained by the separate strong and electroweak theories.

First, there is the striking resemblance among quarks and leptons. Both classes of particles appear fundamental, in that they are structureless at our present limits of resolution. Apart from the fact that quarks carry color but leptons do not, they appear nearly identical. Is this a coincidence, or are quarks and leptons related?

The hint of a connection between quarks and leptons comes from the electroweak theory itself. Unless each lepton family like (e, ν_e) is matched by a color-triplet quark family such as (u, d), the theory will be beset with mathematical inconsistencies. These matched sets are known as quark-lepton generations.

The second puzzling aspect of the theory has to do with the existence of distinct forces of diverse strengths. Here we recall that the electromagnetic interaction, which is of only modest strength between elementary particles, becomes stronger and stronger at short distances. In contrast, the strong interaction becomes increasingly feeble at short distances. Could all the interactions become comparable at some tiny distance, which is to say at some gigantic energy? This would raise the possibility of a common origin for the strong, weak, and electromagnetic interactions. A unification is also suggested by the fact that both QCD and the electroweak theory are gauge theories, with similar mathematical structure.

The strategy for constructing a unified theory is to treat the quarks and leptons symmetrically by joining the quark and lepton families into extended families of fundamental constituents. In the simplest example of a unified theory, each quark and lepton generation is identified with a different extended family. In this way the long-standing mystery of the electric neutrality of stable matter is explained, because the proton and electron must have equal and opposite charges if quarks are combined with leptons in extended families. One branch of the first extended family is the set ($\bar{d}_{red}$ $\bar{d}_{green}$ $\bar{d}_{blue}$ ev_e).

In a gauge theory, each particle in a set can be transformed into any other. Some of these transformations are familiar, such as

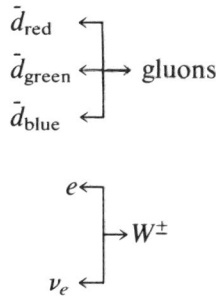

$$\left.\begin{array}{l} \bar{d}_{red} \leftarrow \\ \bar{d}_{green} \leftarrow \\ \bar{d}_{blue} \leftarrow \end{array}\right\} \rightarrow \text{gluons}$$

$$\left.\begin{array}{l} e \leftarrow \\ \\ v_e \leftarrow \end{array}\right\} \rightarrow W^{\pm}$$

but others are novel. The transformations between quarks and leptons, such as

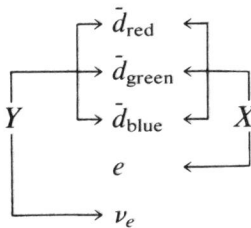

$$\begin{array}{ccc} & \rightarrow \bar{d}_{red} \leftarrow \\ & \rightarrow \bar{d}_{green} \leftarrow \\ Y & \rightarrow \bar{d}_{blue} \leftarrow & X \\ & e \leftarrow \\ & \rightarrow v_e \end{array}$$

can enable protons and bound neutrons to decay. One of the mechanisms for proton decay is shown in Figure 3.17. Here X and Y are hypothetical particles that connect the quarks with the leptons.

In a specific gauge theory, we can compute precisely how the effective interaction strengths change with energy or distance. The evolution of the interaction strengths in the theory is depicted in Figure 3.18. The strong, weak, and electromagnetic interactions strengths are calculated to become comparable at an energy of approximately 10^{15} GeV. The predictions of the theory for the relative strengths of the interactions at current energies agree with precision measurements of

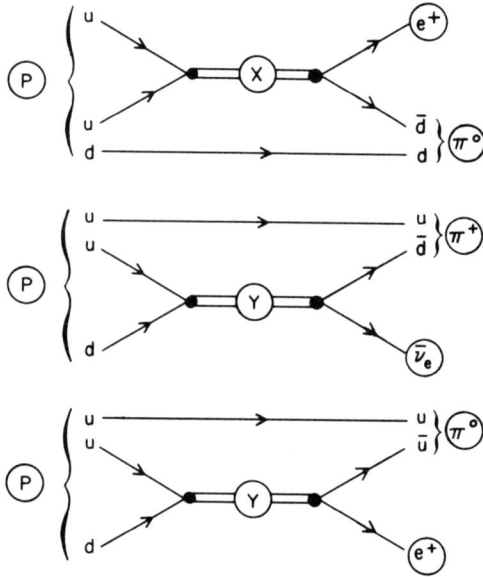

FIGURE 3.17   An example of how a proton might decay in some proposed theories that connect quarks and leptons. In the top diagram, two up quarks unite, leading to the production of a positron and a down antiquark. The down antiquark unites with a down quark to form a neutral pion. Thus in this proposed theory a proton could decay to a positron plus a neutral pion. X and Y are hypothetical particles that connect the quarks and leptons.

the weak interactions and with the masses of the $W^{\pm}$ and Z^0 bosons recently observed. Some unified theories also successfully relate the masses of some quarks and leptons in the same generation, but the meaning of these partial successes is less clear.

The prediction of proton instability is a key consequence of unified theories, and dedicated experiments have been mounted to search for proton decay. The large unification energy implies that the mean proton lifetime must be extraordinarily long—about 10^{30} years or more. Since it is not practical to observe a single proton for such a long period (10^{20} times the age of the universe), it is necessary to monitor extremely large numbers of protons. The largest experiment mounted to date is an instrumented tank of 8000 cubic meters of purified water in a salt mine near Cleveland (Figure 3.19). Currently the searches have yielded negative results that seem to conflict with the predictions of the simplest unified theories.

While the specific prediction of proton decay is a dramatic consequence of unification, the difficulty of studying quark-lepton transitions in the laboratory is apparent. We live in a world in which energies—even in the most powerful accelerators that we can contemplate—are very low compared with the unification scale. However, the discovery of the cosmic microwave background radiation (together with many supporting pieces of evidence) makes it likely that the universe began in a hot big bang of extremely high-energy density. Many aspects of the observed universe find natural explanations in terms of this cosmological model.

Other features of the universe are not so easily understood. Among these, the net baryon number of the universe is of particular interest. The prediction of baryon-number-violating processes, such as proton decay, in unified theories opens the way to understanding why matter dominates over antimatter in the universe.

The phenomenon, technically known as CP violation in K-meson decay, was discovered almost two decades ago. We have no basic explanation for this phenomenon; it may or may not have anything to do with the unified theories discussed in this section. But we should not end this brief survey of what we know in elementary-particle physics without mentioning CP violation. Briefly the phenomenon is as follows. When a K^0 meson, that is, a neutral K meson, is produced, it goes into

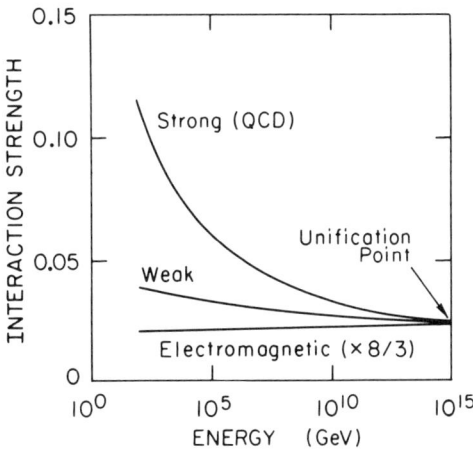

FIGURE 3.18 Example of how in some proposed theories the strong interaction can eventually be combined with the electromagnetic and weak interactions. The idea is that the strength of the interactions depends on the energy at which the interaction occurs. The proposal is that at very high energies, say 10^{15} GeV, all three interactions would have the same strength. This is far beyond any energy that we know how to achieve with present accelerator technology or even with advanced accelerator concepts (see Chapter 5, section on Research on Advanced Concepts for Accelerators and Colliders).

FIGURE 3.19   The Irvine-Michigan-Brookhaven experiment searching for proton decay uses 2000 photomultiplier tubes arrayed in an 8000-cubic-meter tank of very pure
water. The tank is located deep underground in a large chamber of a working salt mine.
A physicist working under water to adjust the tubes wears a special diver's suit to avoid
contaminating the water through contact with his skin.

two different states, called K_L^0 and K_S^0, with different lifetimes. The K_L^0
has the longer lifetime, 500 times that of the K_S^0. The K_S^0 decays through
the weak interaction into two pions. According to a general invariance
principle called CP symmetry, the K_L^0 should never decay into two
pions, but it does, about 0.002 of the time. This is the only reaction or
decay in all of elementary-particle physics where CP symmetry has
been observed to be violated. We do not know what is special about the
decay of the K^0. Perhaps it is an indication of another basic force that
is very weak and is so far manifest only in this decay process; perhaps
it has another explanation. One of the basic principles of relativistic
quantum mechanics is that all physical phenomena must be invariant
under a combination of CP symmetry and time-reversal symmetry.
Therefore this CP violation also represents a violation of time-reversal
symmetry.

4

Elementary-Particle Physics:
What We Want to Know

INTRODUCTION

We saw in Chapters 2 and 3 that developments in elementary-particle physics during the past decade have brought us to a new level in the understanding of fundamental physical laws. This new level of understanding is often called the "standard model" of elementary-particle physics. The establishment of the standard model has brought new maturity to elementary-particle physics, which strengthens its interaction with other areas of physics such as cosmology. Although the standard model provides a framework for describing elementary particles and their fundamental interactions, it is incomplete and inadequate in many respects. As usual, the attainment of a new level of understanding refocuses attention on many old problems that have refused to go away and raises new questions that could not have been asked before.

One measure of the inadequacy of the standard model is the number of basic physical parameters that are required to specify it. At one level, one might accept the existence of certain particles and forces as given *a priori*. Even then, there remain many mysterious inputs, such as the masses of the different particles and the relative strengths of the different forces. At a more fundamental level, one seeks explanations for the choices of elementary-particle species and for the gamut of different fundamental forces.

Thus one may ask how the masses of the different elementary particles are determined—what is the underlying mechanism for mass generation, and how are the individual particle masses related? Why do elementary particles come in sets, or generations, whose individual members have similar masses but different fundamental interactions? Why has this generation structure been copied more than once, and how many copies exist? What is the origin of the overall scale for elementary-particle masses? We know that all the stable matter in the universe is made out of the lightest first generation of elementary particles, while the existence of higher generations might have been essential for the synthesis in the early universe of the matter present in it today. The amount of helium in the universe depends on the number of species of light neutral particles. Stellar evolution and astrophysics would be vastly different if elementary-particle masses were substantially altered. Thus these basic questions about the masses and number of elementary particles bear directly on some of the fundamental aspects of astrophysics and cosmology.

Although the standard model certainly represents a great step forward in the unification of the fundamental interactions, a completely unified framework has yet to be developed. It is natural to suppose that the different strong, weak, and electromagnetic forces known today are simply different manifestations of one underlying force, which may also be related to gravity. Such a grand unified theory would tell us why we have the particular set of force-carrying vector bosons that we know, and why their interactions have such different strengths. Grand unified theories can also tell us why elementary particles like to assemble in the observed generations. In particular, they explain why the electric charges of the electron and proton are simply related, so that conventional matter is electrically neutral. If the electric charges of the electron and proton were not equal and opposite to an accuracy of about 20 decimal places, the electrostatic forces between planets, stars, and galaxies would be stronger than their gravitational forces. Thus any explanation of this equality would be welcome to astrophysicists and cosmologists. They would also welcome the new and exceedingly weak forces expected in some grand unified theories that violate previously sacred physical laws, enabling baryons like the proton to decay. Although the basic principles of such grand unified theories are not necessarily compromised, the simplest examples of such theories make predictions for proton decay that appear to conflict with experiment, and an important question for the future is whether there are alternatives that make testable and successful predictions.

It may well be that none of the above questions has a simple answer when posed at the level of the constituents of matter that currently seem to us to be fundamental. Some physicists believe that the particles that we currently regard as elementary are still so numerous and diverse that they may be composites made up from a smaller and simpler set of more fundamental constituents. Just as our predecessors discovered that the atoms of previous generations can be subdivided into more elementary physical objects—culminating in the recent discovery that protons, neutrons, and other strongly interacting particles are actually made out of quarks—so perhaps we too may discover that quarks and leptons are themselves divisible.

It is possible that free magnetic monopoles (particles containing an unpaired north or south magnetic pole) may exist. They are predicted by some unified theories and may remain as relics of an early stage of the birth of the universe. If they do exist, their masses may be enormous—perhaps 10^{16} times the mass of a proton. Definitive evidence for such monopoles would be extremely important for both elementary-particle physics and astrophysics. In any event, experimentalists must be alert for surprises and unpredicted phenomena. Many of the most exciting and most important discoveries in elementary-particle physics have been the least expected.

It is apparent from this discussion that many fundamental questions are left unanswered, and new ones raised, by the standard model. There is no consensus among elementary-particle physicists as to which of these problems are the most ripe for solution, still less what form any such solution might take. The experimental confirmation of some of the ideas incorporated in the standard model has forced theorists to speculate in many new directions that are not all mutually compatible. Ultimately it will be experiment that has to determine which if any of the different possibilities considered by theorists is the path followed by nature. At the moment, theorists' ideas are insufficiently constrained by experimental realities. Balance can be restored to the science of elementary-particle physics, and a new phenomenological synthesis achieved, only if experiments are soon performed that discriminate among the different physical alternatives. Let us now examine some of these more closely, with a view toward refining our intuition about the most appropriate lines for future experiments.

The Problem of Mass

The elementary-particle masses that are known range between zero and about 100 GeV, as shown in Figure 4.1. Generally accepted gauge symmetries mean that some particles, such as the photon, the gluons,

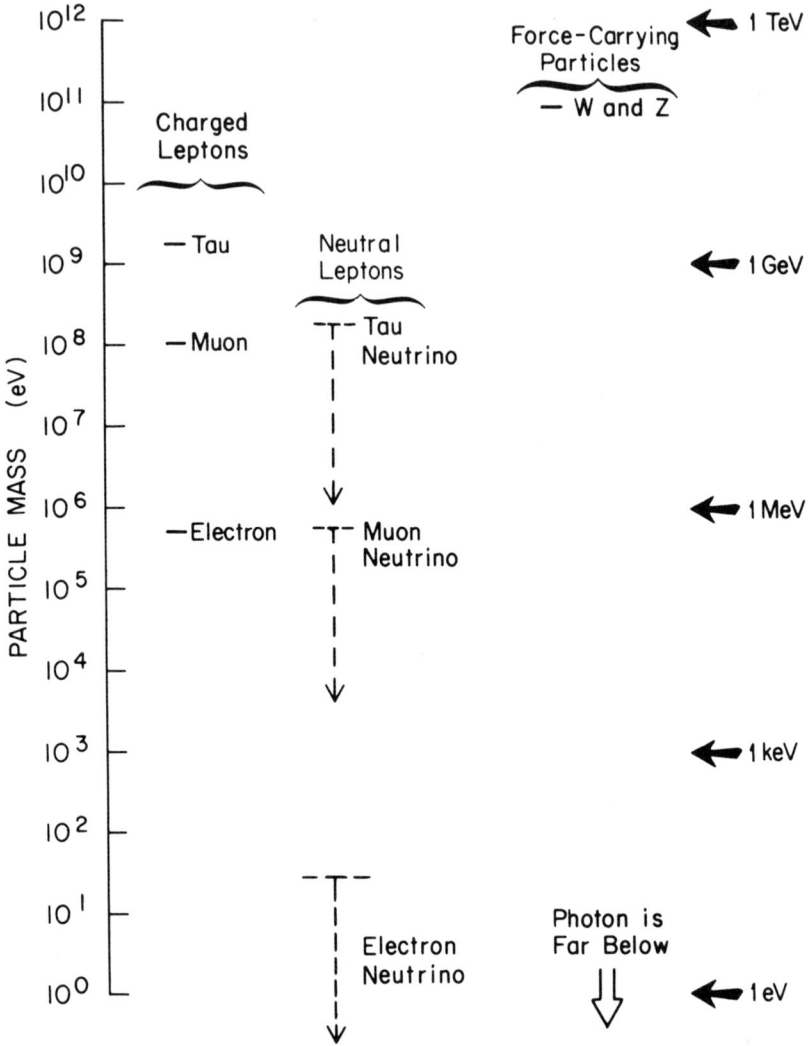

FIGURE 4.1   Some examples of the range of particle masses. The scale extends from 1 eV (1 electron volt) to 10^{12} eV (1,000,000,000,000 electron volts). We are only sure of upper limits on the masses of the neutral leptons or neutrinos. Their masses could be zero. The upper limit on the photon mass is far below the bottom of the page.

and the graviton, are firmly believed to have zero mass. There is no such gauge symmetry to prevent the neutrinos from having masses, although there is as yet no confirmation that any of three known species of neutrino does in fact have a mass. The most stringent experimental upper limit on a neutrino mass is about 10^{-4} of the electron mass for the electron neutrino, and there is an experimental suggestion that it may have a mass just below this limit. There is a much larger mass scale of a different sort associated with gravity, whose extremely weak coupling strength to conventional non-relativistic matter would become strong for matter at a mass or energy of about 10^{19} GeV.

Where Do All These Mass Scales Originate?

Gauge invariance is now part of the theoretical framework of elementary-particle physics, but it forbids masses for all the known particles. For them to acquire masses, gauge invariance must be broken in some way. If desirable features of gauge theories such as their calculability are to be maintained, gauge invariance can only be broken spontaneously. This means that the underlying equations of the theory must possess gauge symmetry, but their solutions need not. This is analogous to the observation that most human beings are not spherical, despite the fact that the laws of physics underlying their construction are themselves rotationally invariant.

The symmetry of a gauge theory will be spontaneously broken if some gauge noninvariant scalar quantity is nonzero in the theory's lowest energy state. Quarks, leptons, and intermediate bosons can then acquire masses in proportion to their couplings to this nonzero scalar quantity. Thus we have a mechanism for generating masses for all the known elementary particles. Unfortunately, gauge theory per se provides little information about the magnitudes of the scalar's couplings to the different quarks and leptons. Thus the wide range of their masses can be accommodated but not explained by gauge theories. To explain their magnitudes we would need an additional dynamical principle.

The original version of the standard model introduced a new elementary scalar particle, called the Higgs particle, to make gauge invariance break down spontaneously. The Higgs particle's couplings to other particles are proportional to their masses, and are hence fixed though unexplained. Clearly it is of vital importance to search for the Higgs particle. Colliding e^+e^- and hadron-hadron beam experiments seem to offer the best prospects, and suitable experiments are envisaged at present and future colliding-beam accelerators.

We note that the ad hoc introduction of a Higgs particle raises new questions. What should its mass be? The standard model provides no answer, and keeping the elementary Higgs mass within acceptable bounds (less than about 1 TeV) proves to be a difficult technical problem.

Composite Quarks and Leptons?

The idea that quarks are the fundamental constituents of strongly interacting nuclear matter was advanced 20 years ago. Since that time this idea has gained universal acceptance, and we know from current experiments that quarks and leptons are structureless, pointlike particles at least down to a scale of 10^{-16} cm. However, the number of these apparently fundamental particles has increased recently to at least 11, not counting the separate red, green, and blue colors for each kind of quark, and also not counting the 11 analogous antiparticles. Thus some physicists are beginning to believe that quarks and leptons may be composites of even more fundamental constituents.

While this hypothesis of another layer to the onion is very seductive, some cautionary remarks are in order. The first is that there are no compelling reasons why any compositeness of quarks and leptons must show up on a scale of 10^{-17} cm, rather than at much smaller and more inaccessible distances. Second, to date there exists no model for composite quarks or leptons that satisfies all the theoretical constraints that such a model should obey. However, our ignorance of a satisfactory model may simply be attributable to a lack of theoretical ingenuity. The only way we shall be able to determine if there is in fact another layer of the onion is by building accelerators that enable experiments to probe distances smaller than those accessible today.

Unification of the Fundamental Forces?

Another persistent theme in physics is the unification of the different particle interactions, the most recent success being the combination of weak and electromagnetic interactions in a unified gauge theory framework. However, the standard model is not completely unified and has three independent gauge couplings. Nevertheless, the underlying gauge principle provides hope that one might be able to find a truly unified theory. One would expect such a theory to make definite predictions for the strengths of all the gauge interactions in the standard model, related to the strength of the underlying unified gauge interaction. This potential unification was described in Chapter 3.

Interaction of Hadrons

So far in this chapter we have been concerned with the properties and interactions of the elementary particles, the quarks and leptons. Although the hadrons are themselves not elementary, we do have a promising theory, quantum chromodynamics (QCD), for the strong interaction of quarks and of hadrons. However, we have not generally been able to apply QCD in a quantitative manner to the interactions of hadrons. These interactions include the dependence of the total interaction probabilities, or cross sections, of hadrons on one another as functions of energy; the elastic scattering of hadrons, in particular at large values of angle or exchanged momentum; the detailed study of lifetimes and decay processes; and the specific production probabilities of hadrons in collision processes as functions of energy and other parameters. One particular class of strong-interaction experiments studies the effects of the spin (intrinsic angular momentum) of hadrons on production and scattering processes. At present we do not know how to use QCD to explain these interactions in detail. We may not be able to do so because the detailed calculations are too difficult to carry out, or because QCD may only be an approximation to the correct theory of the strong interactions.

USING EXISTING ACCELERATORS AND ACCELERATORS UNDER CONSTRUCTION

One of the purposes of this chapter is to set out, in the context of our theoretical understanding, the ongoing program of experimentation at existing accelerators, our expectations for the devices now under construction, and the imperative for major new facilities in the 1990s. For the machines now available we are able to pose many sharp questions. For the machines of the future, the issues are necessarily less specific, but of greater scope. It is, of course, most important to continue to test the standard electroweak theory and QCD and to explore the predictions of unified theories of the strong, weak, and electromagnetic interactions. The degree of current experimental support for these three theories is rather different. For the electroweak theory the task is now to refine precise quantitative tests of detailed predictions. In the case of QCD, most comparisons of theory and experiment are still at the qualitative level, either because a precise theoretical analysis has not been carried out or because of the difficulties of the required measurement. We find ourselves in the curious position of having a plausible theory that we have not been able to exploit in full. So far as unified theories are concerned, we are only beginning to explore their consequences experimentally. Although the simplest model provides an elegant example of how unification might occur, no preferred unified theory has yet been selected by experiment.

Many specific experiments at our existing accelerators will address these issues. Each in its own way, the electron-positron storage rings (SPEAR, DORIS, CESR, PETRA, PEP, and TRISTAN) and the fixed-target proton accelerators (the AGS, the SPS, and the Tevatron) will contribute to the refinement and testing of the standard model. These low-energy tests include the following:

- The study of static properties of hadrons, such as their magnetic moments, charge radii, and masses.
- Studies of polarization effects in hadron physics.
- Further detailed study of the quarkonium states in the ψ and Υ families, with their implications for the force between quarks.
- Investigation of scaling violations in deeply inelastic scattering of electrons, muons, and neutrinos from nuclei.
- Study of the energy dependence of the rate of hadron production in electron-positron annihilations.
- Exploration of how quarks and gluons materialize into hadrons.
- Probing the quark structure of the proton.
- The search for hadrons with unusual composition, such as the quarkless glueball states suggested by QCD.
- Measurement of the rate of dimuon production in hadron collisions, and allied tests of QCD.
- Study of the spectroscopy and decays of states containing c and b quarks.
- Study of the phenomenon of CP violation.
- Searches for rare decays of K mesons to probe for effects of particles perhaps so massive that they cannot be produced at any existing or conceivable accelerator.
- Examination of the interplay of strong and weak interactions in weak decays of one hadron into others.
- Observation of the interactions of neutrinos produced in decays of short-lived hadrons, and demonstration of the existence of the tau's neutrino.
- Refinement of properties of the neutral and charged weak currents.

Many of the experiments listed here are new uses of existing accelerators. In many cases the accelerators were built before the physics of these experiments was known or ever conceived. For example, when the AGS was built, there was little known about K mesons and no conception of glueball states. It was years later that it was realized that the AGS is tremendously useful for searching for the rare decays of K mesons and decades later when it was realized that one could use the AGS to search for glueball states. In general, the elementary-particle physics community has kept old accelerators going only when one could use them for new physics.

We consider next three higher-energy colliders now under construction. Two are electron-positron colliders: the Stanford Linear Collider (SLC) and the LEP facility at CERN. The third collider under construction is the 2-TeV proton-antiproton Tevatron at Fermilab.

SLC and LEP will act as Z^0 factories, as shown in Figure 4.2, yielding studies of the production rate and decay modes of the neutral intermediate boson. Precise measurements of the mass and lifetime of the Z^0 may be confronted with detailed theoretical predictions. This is an important part of the program of probing the electroweak theory in the same way as quantum electrodynamics has been verified. The lifetime and production rate are also measures of the number of quark and lepton species that occur as decay products. This information could provide, among other things, a determination of the cosmologically important number of light-neutrino species. Specific studies of the decays of the Z^0 into heavy quarks will determine the neutral-current interactions of the heavy quarks and also make available a rich source of heavy quarks for the study of their spectroscopy and decays. Some aspects of the strong interactions, including the reliability of QCD calculations and the way in which quarks and gluons materialize into hadrons, will also be explored at the SLC and LEP. It is also conceivable that a light Higgs boson could be observed; it will in any event be important to search for it.

Perhaps the most important work done at the SLC and LEP will be none of the above. Rather, it might be the discovery of another generation of leptons or quarks, or the discovery of a new type of elementary particle, or even the discovery of a new type of force. LEP can eventually produce a higher energy, 200 GeV, than the SLC; hence it will allow exploration to higher energies.

The Tevatron Collider at Fermilab, a 2-TeV proton-antiproton storage ring, will also have a rich and significant physics program. This machine will be a copious source of the charged intermediate bosons W^+ and W^-, whose decays into quarks and leptons define the structure of the weak charged-current interaction. The mass and lifetime of the W are critical parameters of the electroweak theory, like those of the Z^0. Although the Tevatron will not produce as many Z^0's as the SLC or

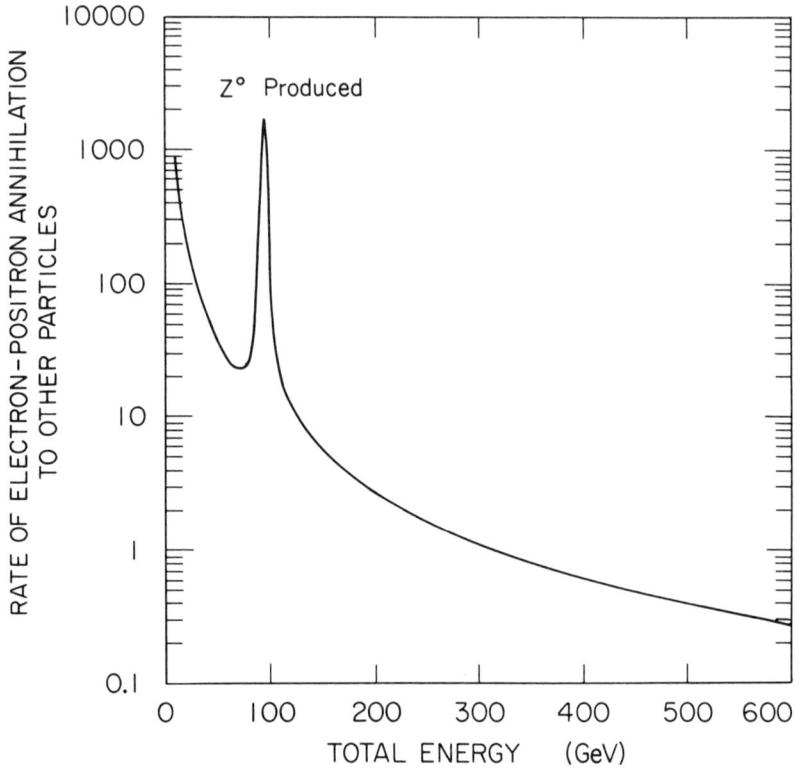

FIGURE 4.2   The rate at which electrons and positrons annihilate to produce other particles is shown as a function of the total energy. In general the rate rapidly decreases as the energy increases. But at about 100 GeV, where the Z^0 particle is produced, the rate has a sharp and useful peak. The electron-positron colliders that will operate at this energy, the SLC and LEP, are called Z^0 factories.

LEP, there should be some systematic advantages to studying both charged and neutral intermediate bosons in the same detector, under similar production conditions. The difference between $W^{\pm}$ and Z^0 masses is a particularly acute probe of the correctness of the electroweak theory. Should there be another intermediate boson in addition to those expected in the standard model, Tevatron experiments would be sensitive to it up to a mass of about 500 GeV. A favorite possibility in theoretical speculations is a right-handed $W^{\pm}$.

Extensive studies will be made of hard collisions among quarks and gluons leading to two or more hadronic jets produced at large angles to the incident beams. This is a superb laboratory for the study of QCD in constituent collisions at energies up to about 600 GeV. Gluon-gluon collisions are quite effective at producing pairs of heavy quarks up to masses of 100 GeV, which would not be accessible in W or Z^0 decays. The discovery of Higgs bosons would also be possible if the mass does not exceed 100 GeV. In any case, the Tevatron represents our first sortie into the several-hundred-GeV regime.

THE NEED FOR HIGHER-ENERGY ACCELERATORS

By early in the 1990s this vigorous experimental program will have subjected QCD and the standard electroweak theory to ever more stringent testing of the kind that is essential to verify that the theories are indeed accurate descriptions of the energy regime below about 100 GeV. Although surprises may well be encountered, it is likely that our efforts to understand why these theories work and to construct more complete descriptions of nature will remain without any direct new experimental guidance. In order to explain what sort of guidance we require, it is useful to summarize some of the shortcomings and open problems of the standard model. Even if we suppose that the ideas of a unified theory of the strong, weak, and electromagnetic interactions are correct, there are several areas in which accomplishments fall short of the announced aspirations, and there are also a number of specific problems to be faced.

• No particular insight has been gained into the pattern of quark and lepton masses or the mixing between different quark and lepton species.

• Although the idea that quarks and leptons should be grouped in generations has gained support, we do not know why generations repeat or how many there are.

• The number of apparently arbitrary parameters needed to specify the theory is 20 or more. This is at odds with our viewpoint, fostered by a history of repeated simplification, that the world should be comprehensible in terms of a few simple laws. Much of the progress represented by gauge theory synthesis is associated with the reduction of ambiguity made possible by a guiding principle.

• CP violation in the weak interaction does not arise gracefully.

• The most serious structural problem is associated with the Higgs sector of the theory. In the standard electroweak theory, the interactions of the Higgs boson are not prescribed by the gauge symmetry as are those of the intermediate bosons. Whereas the masses of the

TABLE 4.1   Questions that Lead to Higher-Energy Accelerators

---

What is the origin of mass?
What sets the masses of the different particles?
Why are there quark and lepton generations?
Are the quarks and leptons truly elementary?
Can the strong and electroweak interactions be unified?
What is the origin of gauge symmetries?
Are there undiscovered fundamental forces?
Are there undiscovered new types of elementary particles?
What is the origin of CP violation?

---

intermediate bosons are specified by the theory, the mass of the Higgs boson is only constrained to lie within the range 7 GeV to 1 TeV. In a unified theory, the problem of the ambiguity of the Higgs sector is heightened by the requirement that there be a dozen orders of magnitude between the masses of W and Z^0 and those of the leptoquark bosons that would mediate proton decay.

• Gravitation is omitted from the quantum theory, although the unification scale for the strong, weak, and electromagnetic interactions is only four orders of magnitude removed from the Planck mass at which gravitational effects become strong. Can gravity be made consistent with quantum theory, and can it be unified with the other fundamental forces?

• Faced with the large number of apparently fundamental quarks and leptons, we may ask whether these particles are truly elementary.

• Are there other types of elementary particles?

• Finally, we may ask what is the origin of the gauge symmetries themselves, why the weak interactions are left-handed, and whether there are new fundamental interactions to be discovered.

Given this list, summarized in Table 4.1, it is not surprising that there are many directions of theoretical speculation that depart from the standard model. Many of these have important implications that cannot yet be tested. Although theoretical speculation and synthesis is valuable and necessary, we cannot advance without new observations. The experimental clues needed to answer questions like those posed above can come from several sources, including

• Experiments at high-energy accelerators;
• Experiments at low-energy accelerators and reactors;
• Nonaccelerator experiments; and
• Deductions from astrophysical measurements.

However, according to our present knowledge of elementary-particle physics, our physical intuition, and our past experience, most of the clues and information will come from experiments at the highest-energy accelerators.

Since many of the questions that we wish to pose are beyond the reach of existing accelerators and those under construction, further progress in the field will depend on our ability to study phenomena at higher energies or, equivalently, on shorter scales of time and distance. What energy scale must we reach, and what sorts of new instruments do we require?

The mystery of symmetry breaking in the electroweak theory, which is to say the nature of the Higgs sector of the theory, presents an especially important and exciting challenge to experimental high-energy physics. This is because there are rather general theoretical reasons why the characteristic scale of the symmetry-breaking phenomenon can be no more than a few TeV. While this probably lies beyond the reach of the current generation of colliders, it is certainly accessible to a hadron machine of multi-TeV capability.

The excitement of the search is heightened by the fact that we know so little of what will be found. Whatever it may be, there is little doubt that further theoretical progress depends critically on finding out. Until we know, the idea of unified theories will rest on a questionable foundation.

Although the Higgs phenomena might possibly occur at less than 1 TeV, building a comprehensive theory in which this occurs proves to be a difficult problem, unless some new physics intervenes.

One solution to the Higgs mass problem involves introducing a complete new set of elementary particles whose spins differ by one-half unit from the known quarks, leptons, and gauge bosons. These postulated new particles are consequences of a new supersymmetry that relates particles of integral and half-integral spin. The conjectured supersymmetric particles stabilize the mass of the Higgs boson at a value below 1 TeV and are likely themselves to have masses less than about 1 TeV. Up to the present, however, there is no experimental evidence for these superpartners.

A second possible solution to the Higgs problem is based on the idea called technicolor that the Higgs boson is not an elementary particle at all but is in reality a composite object made out of elementary constituents analogous to the quarks and leptons. Although they would resemble the usual quarks and leptons, these new constituents would be subject to a new type of strong interaction that would confine them within about 10^{-17} cm. Such new forces could yield new phenomena as rich and diverse as the conventional strong interactions but on an energy scale a thousand times greater—around 1 TeV.

The origin of electroweak symmetry breaking is only one of many puzzles that define the cutting edge of our field. However, because of its importance and accessibility, it imposes a clear minimum requirement on our planning for future facilities. The next high-energy accelerator to be designed and constructed in the United States should be comfortably able to make a few TeV of energy available for new particle production.

Either an electron-positron collider with beams of 1 to 3 TeV or a proton-(anti)proton collider with beams of 5 to 20 TeV would allow an exploration of the TeV region for hard collisions. The higher beam energy required for protons simply reflects the fact that the proton's energy is shared among its quark and gluon constituents. The partitioning of energy among the constituents has been thoroughly studied in experiments on deeply inelastic scattering, so the rate of collisions among constituents of various energies may be calculated with some confidence. As examples, we show in Figure 4.3 how the relative importance of hard gluon-gluon collisions at different energies depends on the energy of the colliding protons. A similar plot for collisions of up quarks and antiup quarks is shown in Figure 4.4.

The physics capabilities of the electron-positron and proton-(anti)proton options are both attractive and somewhat complementary. The hadron machine provides a wider variety of constituent collisions, which allows for a greater diversity of phenomena. The simple initial state of the electron-positron machine represents a considerable measurement advantage. Also, electron-positron collisions give a larger ratio of interesting events to uninteresting background events, and it is easier to find these interesting events. However, the results of the CERN proton-antiproton collider indicate that hard collisions at very high energies are relatively easy to identify. Because the current state of technology favors the hadron collider, it is the instrument of choice for the first exploration of the TeV regime.

A multi-TeV hadron collider will surely reveal much more than the mechanism for electroweak symmetry breaking. Surprises and unexpected insights have always been encountered in each new energy regime, and we confidently expect the same result at TeV energies. Conventional possibilities and existing speculations about the Higgs sector serve the important function of calibrating the discovery reach of a planned facility. They also help to fix the crucial parameters for a new machine: the energy per beam and the rate at which collisions

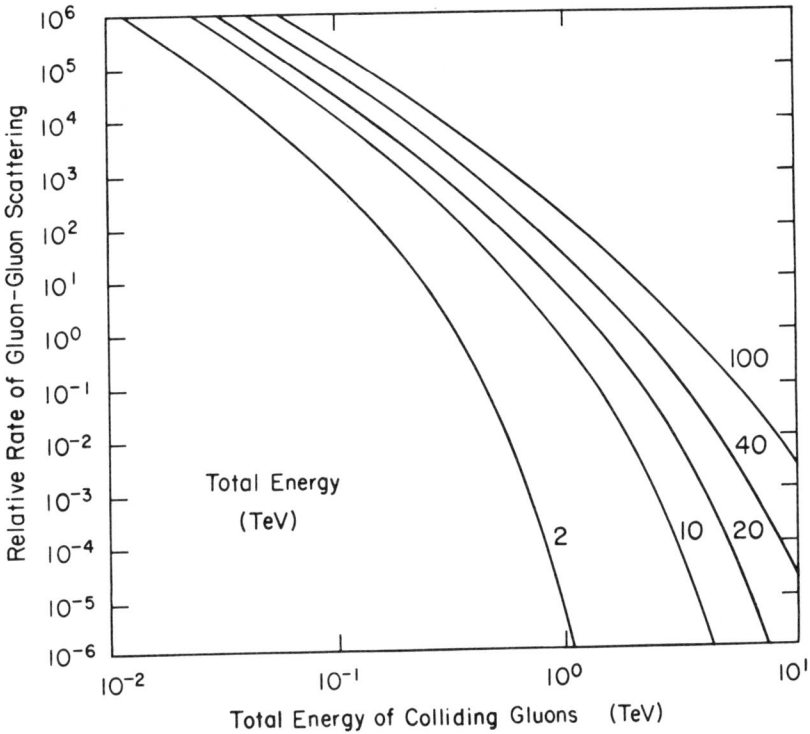

FIGURE 4.3 In high-energy collisions of protons, some of the events actually consist of the collision of gluons within the two protons. Very-high-energy gluon collisions are most interesting. The numbers on each curve give the total energy of the colliding protons; as that energy increases the rate of occurrence of the rare very-high-energy gluon collisions also increases. This is one of the many reasons for wanting to study very-high-energy proton-proton collisions.

occur. Because the most interesting of the anticipated new phenomena are rare occurrences, an ideal storage ring must provide a high collision rate as well as high energies. A total energy of 40 TeV and a collision rate of at least 10^7 interactions per second would allow a thorough exploration of the TeV regime. These parameters define a reasonable target for the next major facility for the study of particle physics in the United States.

Whatever the physics of the TeV energy regime turns out to be, its exploration will provide sorely needed guidance for the attempts at a deeper theoretical description of nature that is now necessarily highly conjectural.

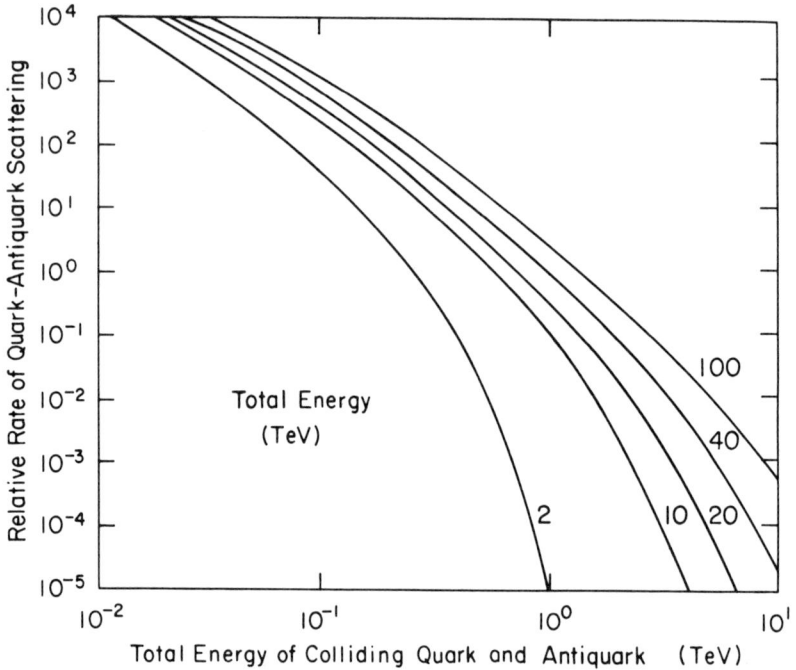

FIGURE 4.4   Quark-antiquark collisions also occur in the collisions of protons. The most interesting quark-antiquark collisions are those that occur at the highest energy.

SOME FUNDAMENTAL ISSUES

It is appropriate to close this chapter with a brief discussion of some fundamental issues for which we do not yet know how to frame a definite experimental program. All the ideas discussed in this report have been formulated within the general framework of quantum field theory. This prescribes that the principles of quantum mechanics be applied locally to fields such as that carrying familiar electromagnetism. A little over a decade ago, there was no such unanimity that quantum field theory was appropriate for describing elementary-particle physics, and many rival approaches were being considered. These have been abandoned since gauge theories have provided such a successful description of the fundamental particles and their interactions. This is not to say that quantum field theory is without its problems.

For example, infinities tend to occur in diagrammatic calculations of the kind described in Chapter 3, but these can be controlled so that computations yield finite and reliable answers. Many physicists have found the existence of even controllable infinities unaesthetic and have sought theories that are completely finite. A class of such theories has recently been discovered, but their relevance to reality is unclear. These theories embody supersymmetry, which has already been mentioned in connection with the Higgs problem, and may aid in the application of quantum principles to gravity.

Unlike the quantization of the electromagnetic field, the gravitational field has never been successfully quantized, and all attempts have ended in a maze of uncontrollable infinities. Some of these infinities are removed by supersymmetry, but others remain. It may well be that the marriage of quantum mechanics and gravitation requires a few more drastic revisions of our ideas. For example, our description of space-time as a continuum may have to be replaced by a discrete, granular structure at extremely short distance. Familiar symmetries such as the equivalence of the laws of nature at all times and places and time-honored conservation laws like the conservation of electric charge may break down in the presence of intense gravitational fields. Perhaps the quantum field theory itself must be rethought or abandoned. Perhaps the usual laws of quantum mechanics should be modified, as has been suggested by some physicists working on quantum gravity.

It does not seem likely that any of these ideas will have a great impact on experimental physics in the near future, but the possibilities should be kept in mind. One of the best laboratories for probing quantum mechanics has been the K^0-$\bar{K}^0$ system studied at high-energy accelerators. Thus even these fundamental problems may have some impact on elementary-particle physics within the next two decades.

5

Accelerators for Elementary-Particle Physics

INTRODUCTION TO ACCELERATORS

The Why and How of Accelerators

Accelerators are the essential tools in most elementary-particle physics research. They provide the high-energy particles used in experiments; the costs of their construction and operation command the major portion of the support budget for particle physics; and a sizable fraction of the community of high-energy physicists is primarily concerned with accelerator technology. Particle physics has always been characterized by the fact that a part of this scientific community has devoted its professional energy and ingenuity to the continuing development of these tools of research. Accelerators thus exemplify imaginative ideas at the frontier of technical complexity and sophistication. The spinoff from accelerator research and development has had applications ranging from radar to controlled thermonuclear fusion and to high-intensity x rays for biological research.

Figure 5.1 shows how an accelerator works. A bunch of electrically charged particles, either electrons or protons, passes through an electric field. The particles gain energy because they are accelerated by the electric field, hence the name accelerator. The energy gained by each particle is given by the voltage across the electric field. Thus an electron passing through a voltage of 1 volt gains an energy of 1

Bunch of
Electrons

Negative High   Positive High
Voltage Plate   Voltage Plate

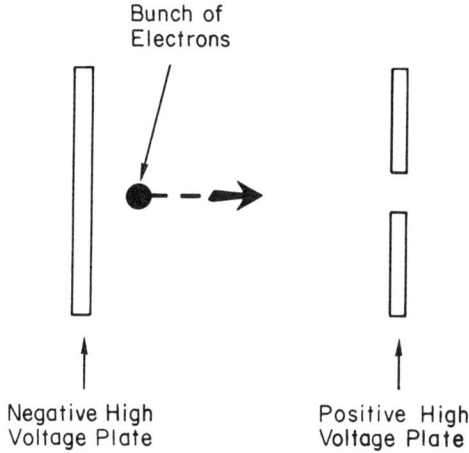

FIGURE 5.1   Accelerators work by exerting an electric force on a charged particle. In the example here a negative plate repels the bunch of electrons and a positive plate attracts them. The electrons thus gain energy in moving from the negative plate to the positive plate. By the time they reach the positive plate they are traveling so quickly that they pass through the hole in the plate and can be used for experiments.

electron volt, abbreviated 1 eV. And an electron passing through 1 million volts gains an energy of 1 million electron volts, abbreviated 1 MeV. In scientific notation 1 MeV $= 10^6$ eV. (Since protons have the same electric charge as electrons, a proton passing through a million volts also gains an energy of 1 MeV.) The highest-energy accelerator in the world is the Tevatron proton accelerator at Fermilab, which is designed to produce an energy of 1 TeV, which is 10^6 MeV or 10^{12} eV.

Accelerators are either linear or circular (Figure 5.2). In the linear accelerator the particle is propelled by strong electromagnetic fields to gain all of its energy in one pass through the machine. In the circular accelerator, the particles are magnetically constrained to circulate many times around a closed path or orbit, and the particle energy is increased on each successive orbit by an accelerating electric field.

Until the 1960s, experiments in particle physics had been conducted using only stationary (fixed) targets. In this case, the beam of accelerated particles is extracted from the accelerator and directed at a fixed target that may consist of a gas, a liquid, or a solid. Usually the target material is the simplest element, hydrogen, whose nucleus is a single proton. A wide variety of proton-proton and electron-proton experiments have been performed that study the absorption or scattering of the beam particles in the target material, the production of new

Linear Accelerator

Circular Accelerator

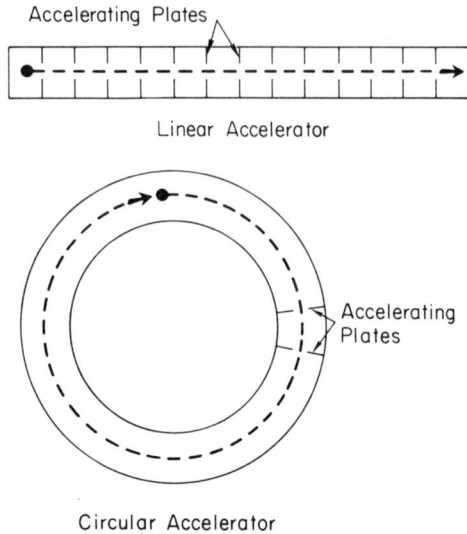

FIGURE 5.2   Very high energies cannot be obtained by using just one pair of plates, as in Figure 5.1. There are two ways to solve this problem. In the linear accelerator, many pairs of plates are lined up, and the particles being accelerated are given more and more energy as they pass through each pair of plates. In a circular accelerator, only one pair of plates is used, but the particles are made to travel in a circle, thus passing through that pair of plates again and again. Each time they pass through the pair of plates they are given more energy.

secondary particles during the collision, and the transformation of the incident and target particles into new kinds of matter.

Not only are the primary reactions of the accelerated particles on fixed targets studied, but also in many experimental situations the secondary particles (such as pions, muons, and *K* mesons) are themselves selected and collimated to produce beams of projectiles that interact with other targets.

As efforts were made to increase the energy in the primary interaction in fixed-target experiments, it was recognized that a large fraction of the energy of the incident particles was not available for the interaction itself but was rather retained as the energy of motion of the recoiling products of the collision. At relativistic energies (i.e., energies that are large compared with the rest energy of the accelerated particles) the collision between a projectile particle and a similar particle at rest makes available for interaction only an amount of energy that is proportional to the square root of the energy of the projectile. That is,

$$E = \sqrt{2mE_{\text{particle}}}.$$

E is the usable or center-of-mass energy, $E_{particle}$ is the energy of the accelerated particle, and m is the mass (in energy units) of the target particle. Thus as the energy of the accelerated particle increases, more and more of it is wasted, since only E is usable. For example, if the energy of the incident particle is increased by a factor of 100, the energy available in the center of mass is increased by only a factor of 10. Eventually it becomes economically and technically impractical to continue to increase the usable energy in fixed-target accelerators. Hence for very high energies we have gone to a different and newer accelerator concept: the particle collider.

Particle Colliders

A simplified example of a particle collider is shown in Figure 5.3. In a circular machine, a bunch of electrons and a bunch of positrons circulate in opposite directions, the particle bunches being held in the machine by a magnetic guide field. (These machines are also called storage rings.) At two opposite places in the machine, the bunches

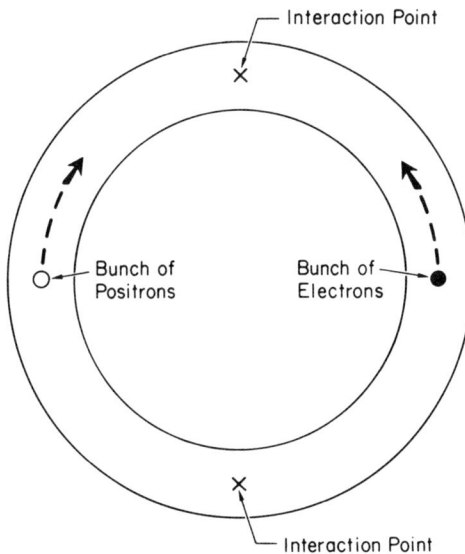

FIGURE 5.3 This colliding-beam storage-ring accelerator has two bunches of particles moving in opposite directions. The bunches collide at the two interaction points. Even though the bunches collide, most of the particles in the bunch pass right through the other bunch; therefore the bunches continue to rotate again and again around the orbits.

collide head on. The usable energy is now

$$E = 2E_{particle},$$

where $E_{particle}$ is the energy of a particle in either bunch. Thus all the particle energy is usable. (This is the usual case, where both colliding particles have the same energy. If that is not the case, as in an electron-proton collider, then not all the particle energy is usable.)

When the bunches come together, most of the particles in one bunch simply pass through the other bunch without actually colliding. Thus they continue to rotate around the storage ring. The bunches may rotate for hours or even days, making thousands or even millions of rotations per second.

The particles are put into the storage ring by an auxiliary accelerator called an injector. In lower-energy storage rings the particles are usually injected with their full energy. In higher-energy storage rings, the particles are accelerated after injection to their full energy. The following combinations of particles are now used or will be used in colliders:

$e^+ - e^-$: electrons colliding with positrons
$p - p$: protons colliding with protons
$p - \bar{p}$: protons colliding with antiprotons
$e^- - p$: electrons colliding with protons
$e^+ - p$: positrons colliding with protons

A critical property of colliders is called luminosity, which is a measure of the rate at which particle collisions occur. Since particle collisions are the essence of particle experiments, the more collisions per second, the more useful the collider. A quantity called the cross section, S, measures the relative probability of two particles colliding. In a collider the rate, R, of collisions per second is

$$R = LS,$$

where L is the collider luminosity. Since the cross section S has units of centimeters squared, the units of L are

$$\frac{\text{collisions}}{\text{centimeters}^2\ \text{second}}.$$

(This is abbreviated as $\text{cm}^{-2}\ \text{s}^{-1}$, the numerator's unit being omitted.) Existing colliders have luminosities in the range of 10^{29} to 10^{32} cm^{-2} s^{-1}.

An alternative to storage rings for particle colliders is the use of colliding beams produced by linear accelerators (Figure 5.4). The colliding bunches of particles pass through each other just once. Much

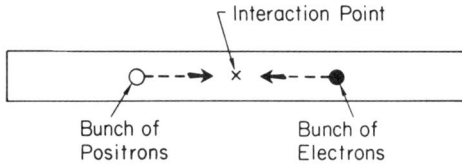

FIGURE 5.4 In this sketch of a linear colliding-beam accelerator the two bunches collide only once. To make full use of that single collision the bunches have to be much denser than in a circular collider.

denser bunches must be used to compensate for the absence of repeated collisions. One form of such a device is currently being constructed that will accelerate in close succession bunches of electrons and positrons in a single linear accelerator. In this case, the charges of opposite sign are separated by magnets and then brought into a head-on collision in a single pass.

Superconducting Magnets in Accelerators

In circular fixed-target accelerators or in circular colliders, the particles are kept moving in a curved path by strong magnetic fields. Those fields are generated by electromagnets that fill most of the circumference of the ring. One of the practical limitations on the achievement of higher energies with circular proton machines has been the size of the ring and the cost of electric power to operate the magnets. The present largest accelerators have a four-mile circumference and consume many tens of megawatts of power.

An innovation that has led to much higher available energy for circular proton accelerators and storage rings has been the development of superconducting magnets. Superconducting metals, such as a niobium-titanium (NbTi) alloy, have zero electrical resistance when cooled to liquid helium temperature. This is a temperature just a few degrees above absolute zero. Since the electrical resistance is zero, no power is consumed in operating electromagnets whose coils are made of a superconductor, although some power must be used for refrigeration to keep the magnets cold. This is one advantage of superconducting magnets.

There is also a second advantage. Superconducting magnet coils can carry extremely high currents. These can give magnetic fields two to four times stronger than ordinary magnets. The circumference of a circular proton machine depends on the strength of the magnetic field for a fixed energy. Hence the use of superconducting magnets allows a smaller circumference to be used or, conversely, a higher energy can be achieved in the same circumference. This has been done during the last several years at Fermilab, where the change from ordinary to superconducting magnets has doubled the energy of the proton accelerator.

Progress in Accelerators and the Energy Frontier

Over the last 50 years there has been a continuous development of new accelerator ideas and engineering achievements. It is remarkable that as each set of concepts appeared to reach a dead end, a new idea, a new technology, has evolved to continue to roll back the frontiers of energy and luminosity. This is most strikingly illustrated in Figure 5.5, which was first published over 20 years ago but is still a good representation of our progress on the energy frontier. Here we have only adjusted our definitions to represent particle colliders in terms of the equivalent energy of the particle striking a stationary target.

ELEMENTARY-PARTICLE PHYSICS AND THE VARIETY OF ACCELERATORS

In the last section we described how accelerators work. We now turn to the reasons for the variety of accelerators used in elementary-particle physics: fixed-target accelerators and particle colliders, proton accelerators and electron accelerators, low-energy accelerators and high-energy accelerators. This variety exists to serve the many different purposes of elementary-particle physics experiments. We will outline these purposes and give some illustrations.

Study of the Properties of Known Particles

Often we know that a particle exists, but we know little about its properties. An example in present-day research is the B meson, which contains a b or bottom quark and has a mass of about 5 GeV. The B meson can decay in many different ways through the weak interaction, and we would like to know much more about these different modes of decay. The cleanest way to study those decay modes at present is to produce a single B meson and a single anti-B meson ($\bar{B}$) in an electron-positron collision using an electron-positron collider. Since the total mass to be created is about 10 GeV, an electron-positron collider that has its maximum luminosity at about 10 GeV is best. Such a collider is the CESR facility at Cornell University. Lower-energy electron-positron colliders do not have enough energy to create the $B\bar{B}$

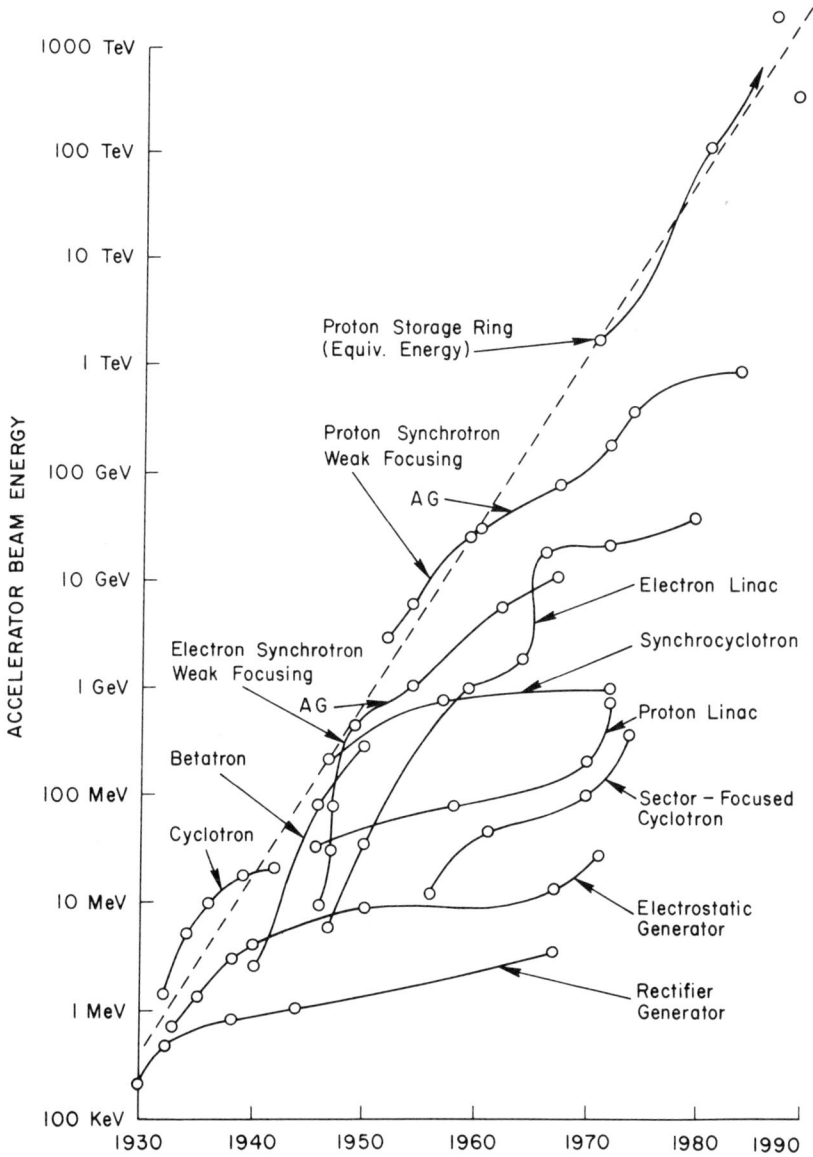

FIGURE 5.5 The maximum energy achievable by an accelerator has increased exponentially with time over the last 50 years. This exponential increase has been maintained by a succession of new inventions in accelerator technology. The highest energies have been achieved by storage rings, the latest invention in accelerator technology. In this figure the energy of storage rings is denoted by the equivalent energy that a fixed-target accelerator would have to possess to give the same useful energy.

pair, while higher-energy colliders have less luminosity at the required energy.

On the other hand, to measure the lifetime of the *B* meson rather than its decay modes, the meson should have high velocity. Then it is best to produce it at higher energy, and the PETRA and PEP electron-positron colliders have that higher energy. Thus the first measurements of the lifetime of the *B* meson were made by experiments at PEP. The recently discovered Z^0 particle is another example. The discovery of the Z^0 was made at the CERN proton-antiproton collider because that was the only existing collider or accelerator with enough energy to create the 93-GeV mass of this particle. But electron-positron collisions should provide the cleanest and easiest way to create Z^0 particles in great numbers so that their properties can be studied in great detail. Indeed, studying the physics of the Z^0 is the first purpose of two electron-positron colliders now under construction, the Stanford Linear Collider (SLC) and LEP at CERN (see the section below on Accelerators We Are Using or Building). Existing electron-positron colliders do not have enough energy to create Z^0 particles.

The study of the decays of *K* mesons provides another example. The puzzling phenomenon of CP violation is observed only in such decays. To study these decays in detail we need a large number of *K* mesons, which are best produced in fixed-target proton accelerators. Thus the Tevatron, the Alternating Gradient Synchrotron (AGS), and the SPS machines are all used to produce *K* beams for various studies of *K*-meson decays.

Study of the Known Forces

Three of the four known forces, the electromagnetic force, the weak force, and the strong force, can be studied using accelerators. But the most suitable accelerator depends on the force to be studied and how it is to be studied. An old but still interesting example is the discovery that the total cross section (that is, the total rate) for the interaction of protons with protons through the strong force increases as the energy increases. The increase is not large, but it is a clear increase. This is called the rising total cross-section effect, and we do not understand why it occurs. To make progress on this problem we need more data on proton-proton interactions at yet higher energy. These data can only come from a higher-energy proton-proton collider.

Further studies of the weak force at higher energy require a different facility. The weak interaction can only be studied in a collision if the strong force is not present; otherwise the strong force masks the weak force. Therefore one of the particles in the collision must be a lepton, because leptons do not feel the strong force. The classic way to study the weak interaction has been to collide neutrinos with protons or with neutrons in a fixed-target experiment. The neutrinos must come from a secondary neutrino beam produced at a proton accelerator.

However, as we discussed in the last section, fixed-target experiments are more limited in their maximum energy than are collider experiments. Thus the highest-energy weak-force studies will have to be done using an electron-proton collider. No such collider exists, but the knowledge and technology needed to build such a facility do exist. The DESY laboratory in Germany is now building such a collider, called HERA.

Tests of New Ideas and Theories

It is rare that a new idea or theory can be tested with experimental data that already exist. More commonly it is necessary to carry out new experiments to test the new ideas or theory. Such experimental tests often stretch the capabilities of the accelerator being used. For example, the principle of lepton conservation states that the decay

$$\text{muon} \rightarrow \text{electron} + \text{photon}$$

cannot occur. This decay has been looked for but has not been found to a precision of about 1 part in 10^{10}. To test some theories that say that this decay should in fact occur at a level of 1 part in 10^{12} the experimenter needs a great number of muons. The best source for such muons is the secondary muon beam from a high-intensity proton accelerator. High intensity, not high energy, is important. Therefore experimenters use a relatively low-energy but high-intensity proton accelerator such as the 800-MeV LAMPF machine at Los Alamos.

Other tests of new ideas and theories require high energies. For example, in Chapter 4 the technicolor theory was mentioned; this theory predicts new particles in the mass range of 1 TeV. No existing collider can produce particles with such a large mass. Therefore a higher-energy proton-proton or proton-antiproton collider is needed. The proposed Superconducting Super Collider (SSC), discussed below in the section on The Superconducting Super Collider, A Very-High-Energy Proton-Proton Collider, would have sufficient energy to produce these massive new particles.

The Search for New Particles and the Mass Scale

The need for higher-energy colliders to search for new particles is so fundamental to our goals that we will discuss this in more detail. There are two questions involved in the search for a new particle. How much energy is needed? How much intensity or luminosity is needed?

The answer to the energy question depends on the type of collider used to produce the particle. In electron-positron colliders, when the electron and positron annihilate, they can give all their energy to the production of the particle. If a single particle is to be produced, then the total energy of the collider need only be equal to the mass of the single particle.

Proton-proton colliders or proton-antiproton colliders require more total energy than the mass of the particle that is to be produced. This is because the production process actually occurs through the collision of a single quark or gluon in one proton with a single quark or gluon in the other proton (or antiproton). On the average a single quark or gluon in a proton only carries about 1/6 of the total energy. Therefore the total collision energy needed to produce a particle of a certain mass is about 6 times that mass. This is a rough rule, because the second question—how much luminosity is required—is also important. If the production process for a new particle is rare, then a high luminosity is required.

The range of masses that can be produced at a collider should overlap the mass range or mass scale of the theory that is to be tested. To achieve this mass range both high energy and high luminosity are necessary. To reach the mass scales of the theories discussed in Chapter 4, colliders should have the following general properties:

proton-proton or proton-antiproton
$\begin{cases} \text{10-TeV minimum total energy} \\\\ 10^{32}\ \text{cm}^{-2}\ \text{s}^{-1}\ \text{minimum luminosity} \end{cases}$

electron-positron
$\begin{cases} \text{1-TeV minimum total energy} \\\\ 10^{32}\ \text{cm}^{-2}\ \text{s}^{-1}\ \text{minimum luminosity} \end{cases}$

As will be described in the section below on The Superconducting

Super Collider, A Very-High-Energy Proton-Proton Collider, we now have the knowledge and experience needed to build a proton-proton collider of 20- to 40-TeV total energy, and the U.S. particle-physics community is now planning for such a project at the upper end of the energy range. On the other hand, we do not yet have the knowledge and experience needed to build an electron-positron collider in the TeV range. The development work being done toward that goal is described below in the section on Research and Development for Very-High-Energy Linear Colliders.

Searches for Clues to Puzzles and Exploration of the Unknown

In Chapter 4 we reviewed some of the puzzles now faced by the particle physicist, such as the origin of mass. The present theoretical ideas intended to solve some of these puzzles may all be wrong. In the end, the experimenter must search for experimental clues to the solution of the puzzles. And even more generally, as in all sciences, the particle physicist simply wants to explore the unknown. The variety of accelerators is needed because no one can be sure of the best methods of searching for clues. For example, suppose it turned out that electrons are not elementary particles after all, but rather are made up of yet smaller or simpler objects. This might show up first as certain strange effects in electron-positron collisions. Or it might be detected first in precise measurements at very high energy of the scattering of electrons on protons in an electron-proton collider. Or perhaps the electron itself might still be elementary, but its heavier relative, the tau lepton, might not be. In that case one would not need very high energies. Instead, the requirements would be copious production and careful study of tau leptons—work that could best be done at a relatively low-energy electron-positron collider.

A dramatic example of exploration of the unknown is the recent idea for the study of a proposed new state of hadronic matter, a quark-gluon plasma. In the usual gaseous plasma, the atoms are ionized, which means that the gas consists of electrically positive ions and negative electrons moving around free of the constraints of being bound together in electrically neutral atoms. Similarly, in a quark-gluon plasma, the quarks and gluons would also move around free of the constraints of being bound together in hadrons. This proposed new state of hadronic matter might be produced by colliding heavy nuclei together in a heavy-ion collider. The ultimate such collider would have uranium colliding with uranium at high energy.

ACCELERATORS WE ARE USING AND BUILDING

In this section we describe the world's high-energy accelerators in some detail. A summary of this information is given in Figures 5.6-5.8, and Table 5.1 briefly describes the current accelerator facilities program in the United States. Appendix A lists accelerators and colliders in operation; Appendix B lists colliders now under construction.

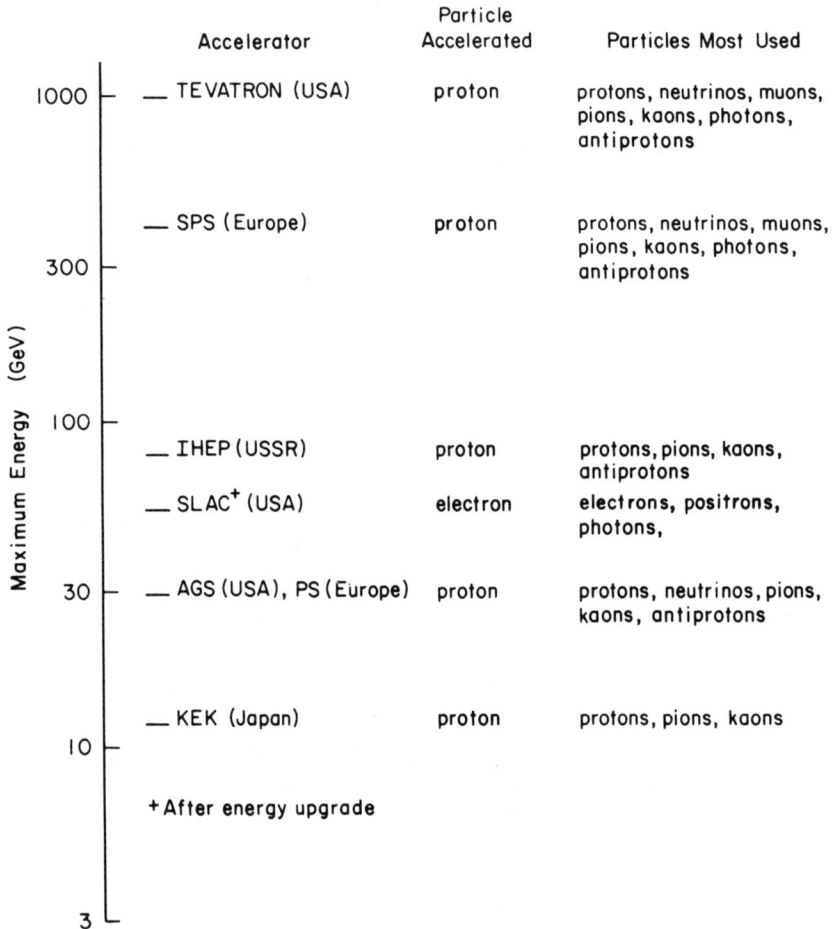

	Accelerator	Particle Accelerated	Particles Most Used
1000	— TEVATRON (USA)	proton	protons, neutrinos, muons, pions, kaons, photons, antiprotons
300	— SPS (Europe)	proton	protons, neutrinos, muons, pions, kaons, photons, antiprotons
100	— IHEP (USSR)	proton	protons, pions, kaons, antiprotons
	— SLAC$^+$ (USA)	electron	electrons, positrons, photons,
30	— AGS (USA), PS (Europe)	proton	protons, neutrinos, pions, kaons, antiprotons
10	— KEK (Japan)	proton	protons, pions, kaons
3			

Maximum Energy (GeV)

+After energy upgrade

FIGURE 5.6   The world's high-energy fixed-target accelerators ordered according to their maximum energy. There are no new fixed-target accelerators under construction, but some, such as the SLAC electron accelerator, are being increased in energy, and the AGS at Brookhaven may be increased in intensity.

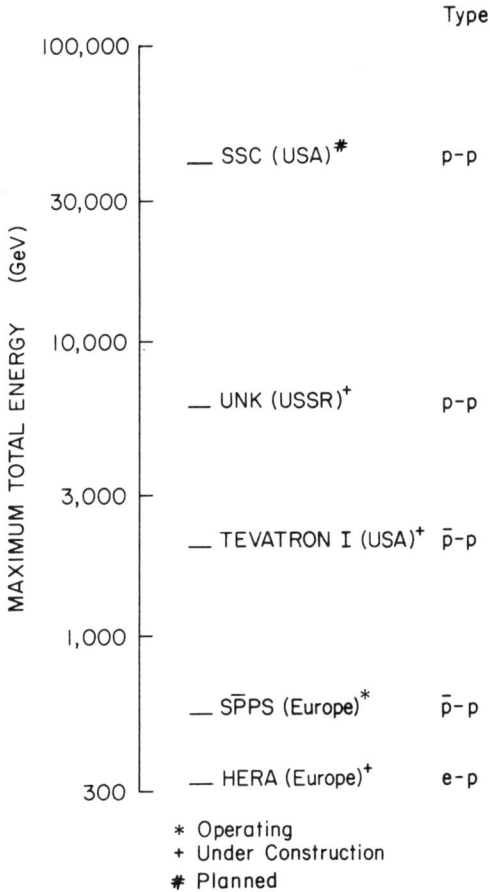

FIGURE 5.7 Proton-proton, antiproton-proton, and electron-proton colliders now in operation or being constructed or designed. The proposal for the highest-energy collider, the SSC in the United States, is now being developed.

Proton Accelerators: Fixed Target

The only high-energy proton accelerators in the United States today are the 30-GeV AGS at Brookhaven National Laboratory and the Tevatron at the Fermi National Accelerator Laboratory, both of which can produce energies up to 1 TeV. The Tevatron, shown in Figure 5.9, is a superconducting proton accelerating ring that was added in the same tunnel with the 400-GeV proton synchrotron that has been in use at Fermilab over the past 10 years. The lower-energy ring, which uses

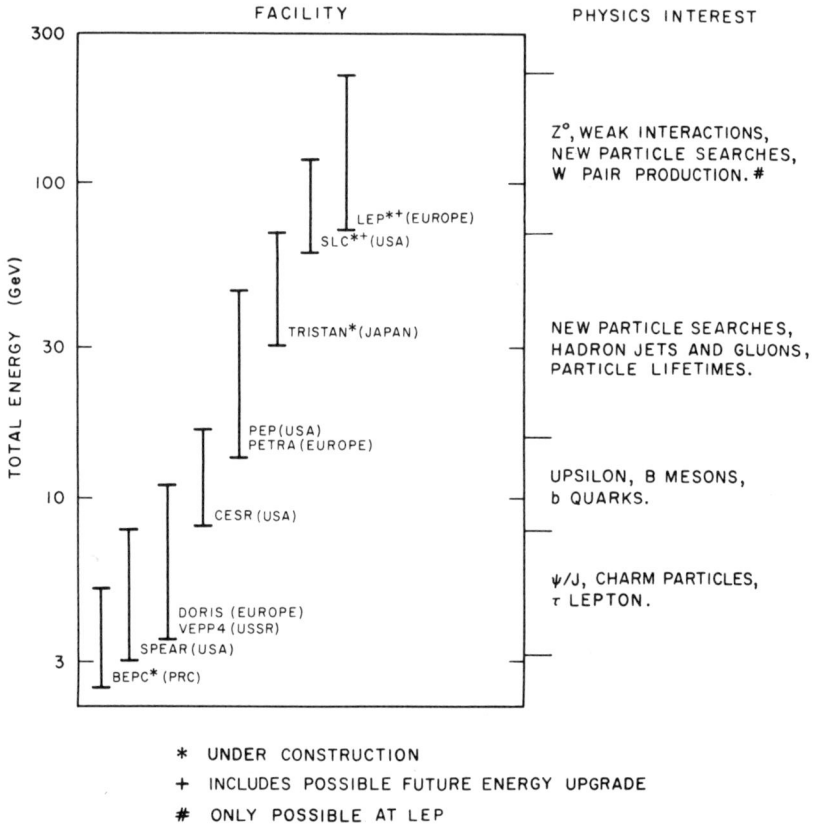

FIGURE 5.8 Electron-positron colliders now in operation or under construction. All these colliders are circular except for the SLC, which is the first linear collider.

conventional magnets, is used as the injector for the new facility. The AGS and the Fermilab 400-GeV machines have their European counterparts in the comparable Proton Synchrotron (PS) and Super Proton Synchrotron (SPS) machines at the CERN laboratory in Geneva, Switzerland. All of these machines have provided extensive data on the systematics of the strong and weak interactions over the past two decades.

The Tevatron will provide for the extension of these experiments into a new energy domain, and, as described later, it will also be used as a proton-antiproton storage ring. Its fixed-target facilities will provide information on the strong-force production of many kinds of hadronic particles. In addition, the decays of these hadrons provide

TABLE 5.1   Current U.S. Accelerator Facilities

Fermilab

The superconducting Tevatron proton accelerator has begun operation, providing proton beams up to 1 TeV for a large variety of fixed-target experiments. The source of antiprotons for the proton-antiproton collider is currently under construction. This facility should be complete in 1986 and will provide 2-TeV total energy, becoming the highest-energy facility in the world.

SLAC

The PEP and SPEAR electron-positron storage rings will continue to operate. However, the main goal is the early completion (1986) of the 100-GeV total-energy electron-positron linear collider, the SLC, providing the first detailed exploration of the exciting physics associated with the Z^0.

Brookhaven

Plans are being made for a major upgrade in the intensity of the AGS proton accelerator and for its improved utilization. The laboratory has recently established a polarized proton facility, and this will be fully exploited.

Cornell

An improvement program has begun for the electron-positron storage ring, CESR, which will increase its luminosity by a factor of 5 or more. Improved detectors will occupy each of the two interaction regions. This program will make possible a much more detailed study of the important upsilon energy region.

extensive and important data on the properties of quarks and on the weak interactions. (Indeed the *b* quark was discovered, and some of the evidence for the *c* quark was obtained, at fixed-target proton machines.)

Both the Tevatron and the older 400-GeV accelerator can produce high-energy beams of muons, neutrinos, and photons. The muon and neutrino beams provide extensive information on the weak force and on the quark structure of protons and neutrons. The photon beams are used to study the photoproduction of various states of hadronic matter.

Although the AGS and the PS are at the lower end of the fixed-target proton accelerator energy range they provide special opportunities to do experiments with intense beams of muons and *K* mesons. Unique beams of lower-energy neutrinos and antiprotons are also provided. In addition, the AGS has a polarized primary proton beam.

Proton-Proton and Proton-Antiproton Colliders at CERN

At the present state of accelerator technology the hadron-hadron collider provides the best mechanism to reach the highest interaction

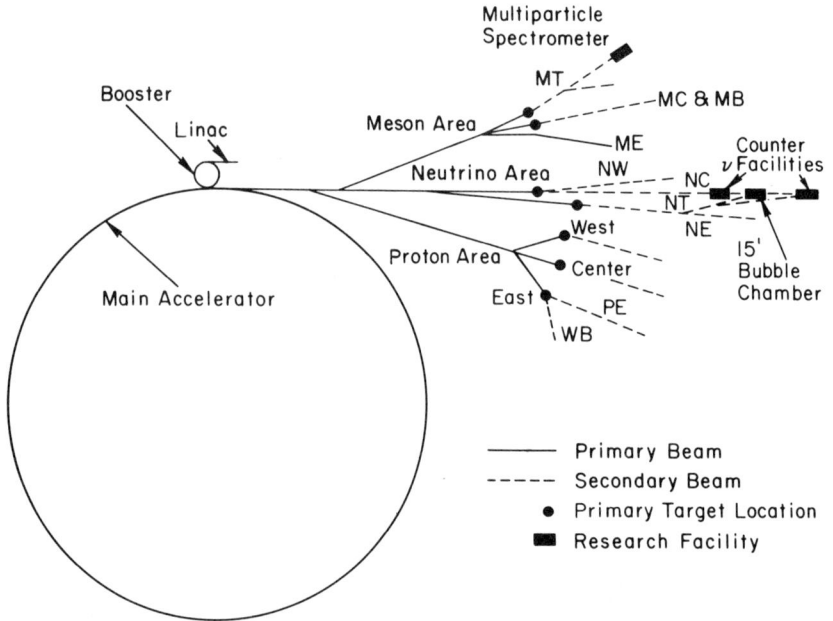

FIGURE 5.9   The 1-TeV fixed-target proton accelerator called the Tevatron at Fermilab is shown here schematically. The large number of secondary beams that can be produced by this accelerator are shown.

energy of the basic constituents of matter. In recent years, these facilities have become important.

The proton-proton collider storage ring ISR (for Intersecting Storage Rings) at CERN provided up to 63 GeV of total energy. It established many aspects of the physics of strong interactions beyond the energies previously available from fixed-target accelerators. This facility was shut down at the end of 1983.

The large CERN proton-antiproton collider ($S\bar{p}pS$) provides about 600-GeV center-of-mass energy by storing counterrotating beams of antiprotons and protons in the main ring of the SPS machine. This is a spectacular feat, since the antiprotons are obtained as secondary beams from another proton accelerator and must be accumulated for many hours before being injected into the SPS ring. This facility provided the mechanism for the long-awaited discoveries in 1982-1983 of the $W^{\pm}$ and Z^0 intermediate-vector bosons of the weak interaction. A remarkable feature of these collisions was that the rare phenomena of the production of the Ws and Zs and their subsequent decay into leptons could be clearly identified and analyzed amid a large background of other reactions.

The 2-TeV Proton-Antiproton Collider at Fermilab

The CERN proton-antiproton collider can study masses up to about 200 GeV. The Fermilab proton-antiproton collider (Tevatron I), which will begin operation for experiments in late 1986, will make it possible to extend studies up to the 600-GeV mass scale. This collider uses the superconducting magnet ring of the Tevatron, with counterrotating beams of 1-TeV protons and antiprotons. Thus a total energy of 2 TeV is obtained. Equally important is the fact that its luminosity should reach 3×10^{30} cm^{-2} s^{-1}. This is critical because the production and study of large-mass particles requires high luminosity as well as high energy. Appendix B lists colliders now under construction.

Thus the Tevatron collider will be able to go well beyond the mass scale accessible to the CERN collider. In addition, it will lay the groundwork for the next step, the construction of a large hadron-hadron collider, which uses superconducting magnets and reaches the multi-TeV mass scale.

Electron Accelerators: Fixed Target

The fixed-target electron accelerators, especially the Stanford Linear Accelerator, have been used to study inelastic and elastic scattering of electrons from protons and neutrons. It was from these experiments in the late 1960s that the first strong evidence for the quark structure of the neutron and proton was obtained. Later, the asymmetry in the scattering of polarized electrons on deuterium at SLAC provided the first definitive evidence of the interference between the electromagnetic and the weak interactions, thus confirming the theory of the unification of the weak and electromagnetic forces. The electron linear accelerator has also been used to provide secondary beams of photons and hadrons for the study of strong interactions and other phenomena such as those described for fixed-target proton accelerators.

Circular Electron-Positron Colliders

Colliding beams of electrons and positrons in e^+e^- storage rings have proven to be an extraordinarily productive accelerator technique. On collision, the e^+ and e^- may produce any particle-antiparticle pair for which there is sufficient energy: an electron-positron pair, a quark-antiquark pair, and so forth. This technique led to the discovery of the ψ (psi) meson (simultaneous with the discovery of the same state with a proton accelerator), the τ (tau) lepton, the D mesons containing a c quark, hadron jets, and most recently the B mesons, which contain a b quark. It has also permitted clean and detailed study of the spectroscopy of the ψ energy states (c plus $\bar{c}$ quarks) and of the Y (upsilon) states (b plus $\bar{b}$ quarks). Because of the absence of other "spectator" quarks (as in experiments with proton beams), these reactions are unusually clean and most amenable to theoretical interpretation.

The SPEAR electron-positron storage ring at Stanford and the CESR electron-positron storage ring at Cornell have uniquely contributed to the great advances in particle physics in the 1970s and early 1980s. The SPEAR ring has provided a wealth of information concerning the ψ resonances, the charmed mesons, and the tau lepton. The CESR facility has powerfully exploited the fact that it is in just the right energy range to study the upsilon resonances, the B mesons, and the properties of the b quark. In Europe, meanwhile, the DORIS facility at DESY has been rebuilt to join in the effort to study the upsilon region. The storage ring VEPP4 in Novosibirsk, USSR, has also joined this effort. China is currently constructing a high-luminosity collider, BEPC, with a maximum energy of 5.6 GeV, which will be well suited to ψ and D meson physics.

The Stanford storage ring PEP and the PETRA machine at DESY in Hamburg are sister machines that operate in the energy range of 20 to 30 GeV and 20 to 46 GeV, respectively. The PETRA machine produced the first firm evidence for the existence of gluons as the particles that carry the strong force. PEP and PETRA are used to study the properties of the c and b quarks and the hadrons that contain them; to study the strong interactions of quarks and gluons, testing the theory of quantum chromodynamics; and to look for new kinds of particles. Recently, experimenters at PEP and PETRA have succeeded in measuring the lifetime of the B meson, a property that is important in understanding the weak interactions of the b quark, as well as the overall relations among the quark generations.

The TRISTAN and LEP Electron-Positron Circular Colliders

At present, there are two new high-energy electron-positron storage rings under construction (Appendix B). The TRISTAN ring in Japan will have a total energy of about 70 GeV. It is hoped that this machine will make possible the observation of the predicted top quark, the last member of the third generation of quarks.

The larger LEP project at CERN, shown in Figure 5.10, is 27 kilometers in circumference and will have a first-stage total energy of

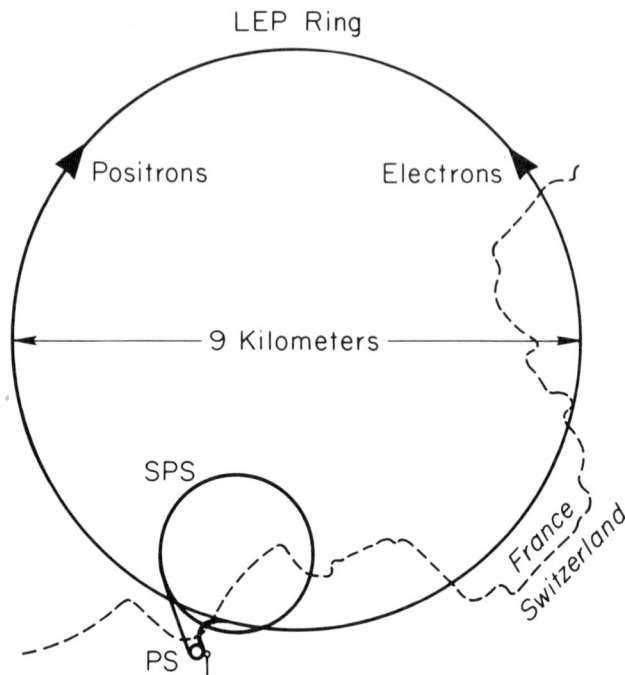

FIGURE 5.10 A schematic illustration of the highest-energy electron-positron collider, LEP, now under construction at CERN.

over 100 GeV. LEP is envisioned as a Z^0 factory that will make possible the careful study of the electroweak interaction and the various decay modes of the Z^0. LEP will be the highest-energy electron-positron collider in the world. Its eventual maximum energy and hence the upper limit on the mass range is somewhat more than 200 GeV. This will provide a clean way to search for new quarks and leptons. At about 160 GeV, we expect the production of pairs of W particles.

LEP and TRISTAN have been designed using existing technology, but it is planned to make use of superconducting radio-frequency cavities at a later date to increase the maximum energy. The development of superconducting accelerator cavities has made encouraging progress at Cornell in the United States and in several laboratories in Europe and Japan. The use of this technology would reduce the operating costs of high-energy electron storage rings and would increase the maximum energy of LEP, for example, to over 200 GeV.

Linear Electron-Positron Colliders

The technology of circular electron-positron colliders has rapidly advanced to a point where performance and cost can be predicted with reasonable confidence. In optimizing the design, the dominant consideration is always the energy lost to synchrotron radiation. This radiation consists of x rays given off by the electrons and positrons as they whirl around the circular ring. The small mass of these particles causes the x rays to be intense and the energy loss to be severe. (This problem is negligible in existing circular proton accelerators and colliders because the proton mass is comparatively large.) The energy lost per turn increases with the fourth power of the beam energy and inversely as the bending radius. This energy must be continuously replaced by a radio-frequency accelerating system, which becomes a major capital investment and major operating expense as the desired beam energy increases. This leads to the rapid increase in the size and cost of electron storage rings as the design beam energy is increased. It is generally agreed that, after optimizing both construction and operating costs, both the size and cost increase with the square of the beam energy.

Several studies of alternative approaches to the design of electron-positron colliders have been made. The conclusion of these studies is that a significant increase in available beam energy will require a new approach that will drastically lower the cost versus energy. The most practical new approach uses two linear accelerators pointing at each other, one accelerating electron bunches and the other accelerating positron bunches. The electrons and positrons move in straight lines, so there is no synchrotron radiation. The problem is in obtaining a useful luminosity or interaction rate in comparison with a storage ring.

This problem has been studied extensively, and technology has now evolved to the point where the luminosity can be achieved by using intense beams of positrons and electrons. At the collision point, the electromagnetic interaction of such intense beams increases the beam density to increase further the interaction probability.

The first linear collider is now under construction at the Stanford Linear Accelerator Center (SLAC), where these ideas have been developed. This Stanford Linear Collider (SLC) has a scheduled completion date of 1986. The SLC uses storage or damping rings to help achieve the high particle densities, and it uses the existing linear accelerator at SLAC to accelerate both electrons and positrons (Figure 5.11). The total energy is 100 GeV, as in the first stage of the LEP storage ring in Europe.

The SLC has two functions: (1) As an elementary-particle physics

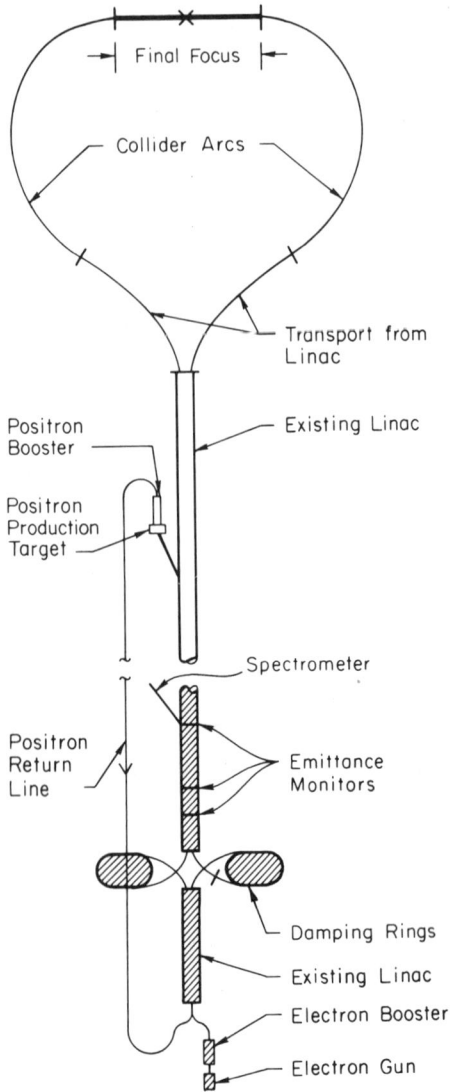

FIGURE 5.11 A schematic illustration of the SLC, the linear electron-positron collider now under construction at Stanford University. This is the first linear electron collider to be built.

facility it will be a Z^0 factory, allowing the detailed study of the physics associated with the Z^0 and its decay modes. It is expected that this machine will come into operation about two years before the LEP storage ring facility at CERN. However, the maximum energy of the SLC is 140 GeV compared with 200 GeV at LEP, and LEP has four interaction regions compared with one at the SLC. The construction cost of LEP is about four or five times that of the SLC. (2) The second function of the SLC is to provide experience with linear collider technology and then to explore the feasibility of electron-positron linear colliders in the TeV energy and mass range.

Electron-Proton Colliders

The DESY Laboratory in Germany has begun the construction of an electron-proton circular collider called HERA with 820-GeV protons colliding with 30-GeV electrons or positrons. The proton ring uses superconducting magnets. The present schedule calls for initial operation in 1990. The design calls for some important features for such a facility. A luminosity of almost 10^{32} cm^{-2} s^{-1} is expected, and the electron and positron beams can have longitudinal polarization. This type of polarization is important in the study of the effects of the weak interaction.

The TRISTAN electron-positron collider now under construction in Japan can also be extended to be an electron-proton collider. This would be done by adding an intersecting proton storage ring; a complete design for this possible addition has been made.

Both the circular and linear collider principles can be used to make electron-proton colliders. In the future, large circular colliders will almost certainly be built initially as proton-proton or proton-antiproton colliders. Then, if the physics warrants, an intersecting electron storage ring can be added. Conversely, very-high-energy electron-positron colliders will probably be built in the linear collider form. Then, in a later upgrade, one of the linear accelerators can be modified to accelerate protons, or an additional proton accelerator can be added.

THE SUPERCONDUCTING SUPER COLLIDER, A VERY-HIGH-ENERGY PROTON-PROTON COLLIDER

Physics Goals

As described in Chapter 4, we are now faced with a set of fundamental questions that require experiments to be carried out in the mass

range of several TeV:

• What is the origin of mass, and what sets the masses of the different elementary particles? Is the Higgs hypothesis correct, and can the Higgs particles be found? If the Higgs hypothesis is wrong, what replaces it?

• Are there more quark or lepton generations? Why do these particles form generations?

• Are the quarks and leptons truly elementary?

• Are new theoretical ideas like technicolor or supersymmetry correct? Can the strong and electroweak interactions be unified?

• Are there undiscovered fundamental forces?

The mass range needed to study these problems is illustrated in Figure 5.12. This mass range cannot be reached with fixed-target accelerators; it requires a hadron-hadron collider. As mentioned earlier, when hadrons collide, the full energy of the hadrons is not available for conversion into mass, even in a colliding-beam accelerator. This is because the hadron-hadron collision really consists of a quark-quark, quark-gluon, or gluon-gluon collision; and these constituents only carry a fraction of the total energy of the hadron. The rough rule is that 1/6 of the total energy is available, on the average, for conversion into large masses. We emphasize that this is an average. There is a large probability that 1/10 to 1/20 of the energy can be converted into large masses and a small probability that 1/3 can be converted.

Collider Goals

These physics goals, searching for answers to fundamental questions and exploring new physics in the several-TeV mass range, require a hadron-hadron collider of very high energy and large luminosity. Our knowledge and experience in accelerator technology enables us to set the practical goals for the collider of a maximum energy of 40 TeV and a maximum luminosity of 10^{33} cm^{-2} s^{-1}. To achieve this luminosity a proton-proton collider is favored. The richness and range of the particle physics that can be done at such a facility dictates that there be multiple interaction regions for particle detectors. Six or more interaction regions are desirable. Summarizing, the practical goals for the Superconducting Super Collider are as follows:

Maximum total energy	40 TeV
Maximum luminosity	10^{33} cm^{-2} s^{-1}
Number of interaction regions	6 or more

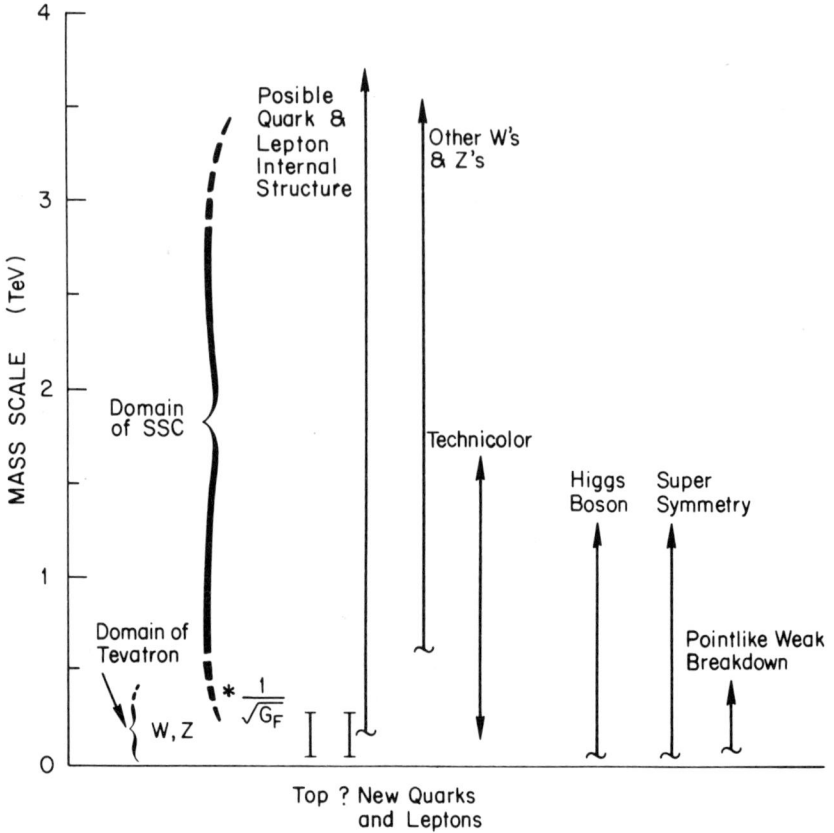

FIGURE 5.12   The mass scale at which physicists believe that a number of fundamental new phenomena may appear. The SSC would extend this scale beyond the mass of a few tenths of a TeV that can be probed by facilities now under construction to the regime of 2 to 3 TeV and above.

Design Studies

Since 1982 the U.S. elementary-particle physics community has been developing a plan for the construction of a high-luminosity proton-proton collider in the energy range of 40 TeV. The work began in the summer of 1982 at a meeting in Snowmass, Colorado (see *Proceedings of the 1982 Division of Plasma and Fluids Summer Study on Elementary Particle Physics and Future Facilities*, June 28-July 16, 1982, Snowmass, Colorado, R. Donaldson, R. Gustafson, and F. Paige, eds.). The result of this and other studies was that a recommendation for the construction of such a facility was made to the U.S. Department of Energy by the 1983 High Energy Physics Advisory Panel of the Department of Energy (HEPAP) Subpanel on New Facilities. This collider has been named the Superconducting Super Collider (SSC) because it requires the use of superconducting magnets to keep its size and its operating power costs within reasonable bounds. Although the basic technology for the machine is at hand, the scale is unprecedented. Therefore an intensive series of design studies has been carried out.

In April 1983 an informal one-week workshop (see *Report of the 20 TeV Hadron Collider Technical Workshop*, Newman Laboratory, Cornell University, Ithaca, New York) was held at Cornell to study the design problems and to make initial estimates of feasibility, time scale, and costs. This was followed by meetings and workshops on hadron collider detectors, on the physics that can be done at the SSC, on accelerator issues related to the SSC, and on cryogenic issues related to superconducting magnets for accelerators. During this period a subpanel of HEPAP was set up to provide advice on the content and implementation of a preliminary research and development (R&D) effort. The most intensive design work at present is the National SSC Reference Designs Study (see *SSC Reference Designs Study Group Report*, May 1984), which was conducted from February through May 1984. This study addressed three areas:

• Technical feasibility: the designs of 40-TeV total-energy proton-proton colliders were explored using three of several possible superconducting magnet styles as study models.
• Economic feasibility: the likely cost range was estimated using preliminary engineering designs for the three magnet styles and the other hardware and conventional facilities required to construct and operate technically feasible colliders.
• Required R&D: the R&D needed to verify design calculations and technical assumptions was identified.

It was not intended, however, that the *Reference Designs Study Group Report* be either a design proposal or a site preference study. Some of the material in this section is based on this study.

Superconducting Magnets

The feasibility of constructing the SSC has been substantially enhanced by the recent success in accelerating protons to high energy in a superconducting accelerator, the Tevatron at Fermilab. This machine uses about 1000 superconducting magnets in a circumference of about 4 miles. It now operates at 800 GeV for physics experiments, and it has also operated in the beam-storage mode preliminary to its use as a proton-antiproton collider. Although the full operating energy of 1000 GeV and full beam intensity are yet to be attained, the performance of the Fermilab machine is a definitive verification of the practicality of using superconducting technology for obtaining beams of very high energy. The Tevatron has opened the door to a new era.

In the course of its construction a great deal has been learned. Great strides have been made in the development of superconducting, niobium-titanium cables, so that high-quality materials are now available in large quantities at reasonable cost. The technique for constraining the superconducting cable in the magnet with the required high precision has been well demonstrated, and protection systems have been developed to cope with the inevitable magnet quenches. (A superconducting magnet quenches when a part of its coil becomes too warm to maintain zero electrical resistance.) During the same period, the industrial capability for producing refrigeration equipment has grown rapidly, and large machines with much higher reliability are now available. The ability to transport large volumes of helium liquid over long distances has been demonstrated. Automatic control over the refrigeration and cryogenic systems has been remarkably successful, and the capability for beam location and control has been demonstrated.

Preliminary Collider Designs and Considerations

The design studies, particularly the National SSC Reference Designs Study, have shown that a conservative extension of existing or near-term technology can lead to the successful achievement of an SSC. Several design options exist, and the selection of a particular design to optimize the cost is one of the most important considerations. The final cost will depend on the results of the R&D program that will be carried out before initiating construction. One of the principal factors determining the detailed design of the collider is the strength of the magnetic guide field. The options cover a broad range of magnetic-field values. The Reference Designs Study has considered the three superconducting, niobium-titanium magnet designs (a), (b), and (c) listed next. Other work has considered the design (d). As shown in Figure 5.13(a), the diameter of the collider decreases as the magnetic field increases.

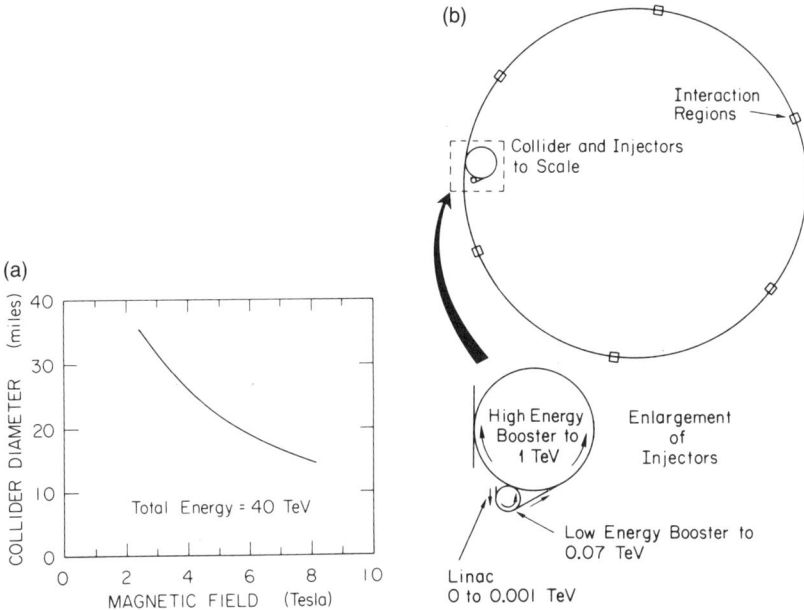

FIGURE 5.13 (a) The diameter of a proton-proton or proton-antiproton collider depends on the total energy desired and the magnetic field used. (b) Schematic layout of the SSC, indicating the injector complex and the main ring, where protons are accelerated to 20 TeV in counterrotating bunches that collide at six points around the circumference. The total collision energy is 40 TeV.

(a) A high-field magnet design has a 6.5-tesla field, with both beam tubes and both coil sets side by side in a common iron yoke contained in a single cryostat. This approach is referred to as the 2-in-1 design. Intrinsic to this approach is magnetic coupling, limiting the extent to which the field strengths in the two apertures may differ. This results in an SSC main ring about 18 miles in diameter.

(b) A medium-field dipole magnet has a 5-tesla field, with each beam tube and coil in its own cryostat. Each cryostat has only enough iron to shield one coil from the magnetic field of the other. This is referred to as the 1-in-1 no-iron design. This results in an SSC main ring about 22 miles in diameter.

(c) A low-field magnet has a 3-tesla field, with each beam tube and each coil set in separate iron yokes, one above the other, in a single cryostat. In this design, although the iron is driven well into saturation, the field is determined primarily by the iron pole faces, and the magnetic fields of the two rings are not strongly coupled. This magnet is referred to as the superferric design. This results in an SSC main ring about 32 miles in diameter.

(d) The designs listed above use a niobium-titanium superconductor, with which we have a great deal of experience. Very high magnetic fields, 8 teslas or more, can be achieved with a niobium-tin superconductor, but there is little experience at present with such magnets.

The choice among these systems is complex. The medium-field technology and to some extent the high-field technology have already been proven in the Tevatron and in the Brookhaven CBA design. Although such magnets could simply be copied and manufactured in quantity, without cost-saving design and production changes the overall cost of the installation would be great. The accelerator tunnel in this case would have a moderate length.

The low-field design is expected to be reliable because of the low values of the forces and low field strengths in the superconductor. It also has the advantage that since the iron profile largely determines the field accuracy, it should be less sensitive to the placement of conductors. It has the disadvantage of requiring a larger tunnel perimeter.

The use of very high field magnets would minimize the tunnel length. However, suitable superconducting cable has not yet been produced in quantity, and the appropriate technology has not yet been developed. This type of magnet might therefore require much more R&D than lower-field designs.

No matter what magnetic-field strength is chosen, the cost of the collider can be reduced if magnets with a smaller aperture can be developed and used. This requires R&D both in magnet design and in the accelerator physics of the collider. Finally, advantage must be taken of the increased scale of production. New fabrication methods suited to mass production will have to be developed.

The SSC facility is shown schematically in Figure 5.13(b). The collider itself sets the size of the site. The injector complex would lie against one portion of the collider ring. The six interaction regions would be distributed around the ring.

An important question is the site required for such a machine. A large number of factors must be taken into account in the site selection. These include ring diameter; environmental considerations; availability of water, power, and roads; and proximity to airports, villages, and cities. Starting first with the technical considerations, it is clear that the number of suitable sites will be strongly dependent on the radius of the machine. From the point of view of beam dynamics, gentle deviations from flatness of the ring might be tolerated. One might be able to take advantage of this in order to locate the interaction halls and service buildings near the surface; and it may permit the use of contour-following, cut-and-cover techniques for the machine closure instead of the more expensive mode of tunneling. A cursory search for suitable sites has suggested that several can be found that would be suitable for even the largest of the rings, one of 100-mile perimeter or more.

Schedule and Cost

Research and development will be needed before beginning the construction of the collider. About 2 years will be required before a working design can be established. It will be necessary to learn how to mass produce low-cost, high-quality magnets and how to handle, mount, and survey the magnets into position with a high degree of precision. It is likely that a full-scale prototype of a relatively long tunnel section and guide field will be constructed in order to test the practicality and integration of the system. It may even be necessary to work on the design of more than one of these systems in parallel in order to determine the minimum-cost system. Such R&D activity is essential to carry out the design.

The scale of this project far exceeds any of our existing high-energy physics facilities. It is obvious that an administrative organization will be required that is responsible to a broadly based national representation of the elementary-particle physics community. The federal funding agencies must indicate that they are receptive to a proposal to build such a machine. International cooperation with respect to building some of the detectors or other costs should be explored.

The Reference Designs Study has considered the construction schedule, as follows: "In this study, we have assumed a six-year construction period, which would lead to completion in early 1994 if construction were to begin in FY 1988. The optimum duration of the construction period should itself be an object of study. . . It will depend on many factors, such as the detailed scope of the facility that is ultimately proposed, the technical means devised for its construction, and the spending pattern needed. Finding ways for minimizing the delay between start of construction and first use for physics research must be given great emphasis."

The same study has estimated the construction costs of the SSC. These costs, based on the three magnet technologies (a), (b), and (c) listed above, range from $2.70 billion to $3.05 billion in fiscal year 1984 dollars. (The costs of research equipment, preconstruction R&D, and possible site acquisition are not included.) Quoting the study, "The contingencies are intended to be sufficiently conservative that these totals represent our best estimate today for an upper bound on the SSC cost. With intense R&D and effective planning, lower costs could result."

RESEARCH AND DEVELOPMENT FOR VERY-HIGH-ENERGY LINEAR COLLIDERS

Physics Motivation

In the section above on Elementary-Particle Physics and the Variety of Accelerators we saw that hadron-hadron and electron-positron colliders largely complement each other in the physics that they explore. As mentioned earlier, there is a rule of thumb that an electron-positron collision has the same *available* energy as a proton-proton collision when the actual energy of the electron plus positron is about 1/6 of the actual energy of the two protons. Thus to reach the same available energy as the planned 40-TeV proton-proton collider, an electron-positron collider would require a total energy in the several-TeV range. This is beyond the reach of the known technology of circular electron-positron colliders; thus a new electron-positron collider technology such as the linear collider is needed.

Incidentally, although most of the thought and work on linear colliders is for electron-positron machines, the concept may also be applicable to electron-proton colliders.

Present Technology and Concepts

As described above in the section on Accelerators We Are Using and Building, the first application of linear collider principles is now being made in the construction at SLAC of the Stanford Linear Collider, a facility with a maximum total energy of 100 to 140 GeV. Starting from this machine, we now consider what R&D is needed in order to build a much larger TeV machine. In linear accelerators and colliders, the critical parameter is the accelerating gradient, i.e., the energy gained per meter of length. In the SLC, it will be about 20 GeV per kilometer.

A 2-TeV collider based on the present SLAC accelerating structure would consist of two conventional linear accelerators each 50 kilometers in length. With 12 electron-positron bunches per pulse, there could be a magnetic switchyard that would feed the bunches to 6 parallel interaction regions, each with a luminosity of the order of 10^{32} cm^{-2} s^{-1}. Using the electrical efficiency of today's pulsed radio-frequency power sources, the total power consumed would be approximately 300 MW. These numbers are quite large. It is desirable to reduce the length and hence the construction cost of such a machine, and also its power consumption. Research and development work aimed toward these goals is now beginning. Of course experience with the operation of SLC will also stimulate progress toward these goals.

One of the directions for improving the technology of linear colliders is to reduce the wave length (increase the operating frequency) of the accelerator structure. A reduction to 5 cm (SLAC uses 10 cm) doubles the accelerating gradient and doubles the electrical efficiency. Research and development is required to produce high-power klystrons at this higher frequency, and several ideas exist for ultra-relativistic or laser-driven klystrons that could provide not only the requisite power but also much higher electrical efficiency. Alternative accelerator structures that promise much higher accelerating gradients will also be explored. (At these shorter wavelengths there is increased energy spread in the accelerated beam, and further development in chromatic corrections of the final focusing systems is required to handle this energy spread.)

The repetitive nature of linear accelerators naturally suggests automated production techniques to reduce construction costs. Also, energy-recovery schemes, perhaps using superconducting microwave accelerator units, need to be explored to increase overall electrical efficiency further. As these technologies advance, the design of a linear collider facility can be optimized, and the construction and operating costs can be reduced.

RESEARCH ON ADVANCED CONCEPTS FOR ACCELERATORS AND COLLIDERS

To conclude this chapter we discuss some advanced ideas for accelerators and colliders. We do not know if any of these ideas can be reduced to practice. But if we are to move substantially beyond the energy range of present accelerator technologies, we must find new ways to accelerate particles. This section describes some of the ideas now being explored.

Linear Accelerators and Colliders

Calculations and research are being carried out in the United States and abroad on a variety of new and advanced concepts for obtaining higher accelerating gradients, which is the energy gain per unit of accelerator length. There is reason to believe that accelerator structures can be built to handle up to 200 GeV per kilometer, ten times the currently available gradients. What is needed is a suitable high-efficiency, high-power source of short-wavelength electromagnetic radiation that can provide a relatively large amount of energy per unit length. We list some of the possibilities:

1. Very-high-power, very-short-pulse-length, high-frequency klystrons suitable for this purpose may be developed.

2. A special case of a source of short-wavelength radiation is the wake field of a high-energy beam passing through a cavity system. This idea is being pursued theoretically and shows considerable promise and a special simplicity since the wake-field source cavity can be combined with a beam-accelerating cavity within a single structure.

3. In the two-beam accelerator concept, a high-power, low-energy electron beam travels parallel to the desired high-energy particle beam. Using a principle such as that of the free-electron laser, the high-power, low-energy beam radiates its power to the high-energy beam, thus providing the acceleration.

4. A more radical approach is to use the very short wavelength obtainable from a laser. In this case one cannot consider accelerating structures of conventional design; the dimensions are far too small. It appears possible, however, to use a suitable optical grating in place of a conventional cavity. The most extreme case would be obtained if the periodic grating were replaced by a periodic plasma, possibly formed over a grating surface. In this case gradients as high as 1 TeV per kilometer could theoretically be attained. Such high and obviously desirable gradients can only exist in or near a plasma and not in or near any solid conductor or dielectric.

5. A particularly interesting solution occurs when a plasma is exposed to two laser beams of suitably close frequency. The beat frequency between the two lasers can be matched to the natural plasma frequency, and a strong periodic and moving charge modulation can be induced. Large electrostatic fields are generated by this modulation, and these could be used to accelerate suitably injected beams. Accelerating fields as high as 2 TeV per kilometer have been discussed, but there remains great uncertainty about the stability, energy efficiency, and suitability of such a mechanism to the construction of a high-energy linear collider.

Many such ideas have been suggested. Some of them may not work. Others may work but not have application for high-energy physics. It is clear, however, that without some such idea, no great further step in energy will be possible. On the other hand, with gradients of the order of 1 TeV per kilometer theoretically possible, an accelerator of 100 TeV is not unthinkable. It is thus important to the future of the field that these ideas are followed up.

Ultrahigh-Energy Circular Colliders

We have a great deal of knowledge and experience with the technology to be used to build a 40-TeV proton-proton circular collider. The primary limitation of that technology is that we do not know how to increase substantially the magnetic field that guides the particles in a circle, and hence we do not know how to decrease substantially the circumference of the collider. Some size and cost reduction can be obtained in guide-field magnets by the use of new superconducting materials such as niobium-tin. While such developments are important, they do not promise a radical saving or access to much higher energies. Mechanical forces will limit the usable magnetic fields no matter what conductors become available.

Even if we could substantially decrease the circumference of a proton-proton collider, we would then reach a second limitation: the protons would begin to lose large amounts of energy via synchroton radiation, as occurs at much lower energies in circular electron-positron colliders. Indeed, no ideas have yet been proposed to enable an increase of the energy of a circular collider beyond the 100-TeV range.

The Need for Advanced Research on Accelerators and Colliders

Thus new accelerator ideas need to be developed and explored. In the past, new ideas have indeed occurred, resulting in the enormous increases in accelerator energy that have been achieved in the past 50 years. However, the present scale of R&D in accelerator technology is small and certainly not commensurate with its importance. Part of the problem is the reluctance of individuals to commit themselves to tasks whose possible fruition seems quite distant. Another problem is the lack of suitably trained multidisciplinary experts. A third may be traced to the mechanisms for supporting accelerator physics. Encouragement to universities to expand training in accelerator physics is needed. Possibly, too, it would be desirable to have a funding mechanism that would allow laboratories to pursue such work with an assurance that such funding was truly an addition to that for more immediate goals. There is a strong and natural tendency for internal priorities to cut back on such long-range activities.

Despite these reservations, it is encouraging to note that there are still many people working on new ideas and that advanced accelerator workshops and schools take place regularly. We can hope and expect to see significant new activity in the coming decade.

6

Instruments and Detectors for Elementary-Particle Physics

INTRODUCTION

Elementary particles cannot be seen directly; their path and energy as they come out of an accelerator or out of a collision must be determined indirectly. It is also important to identify the type of particle: electron, muon, proton, or photon, for example. Thus the detectors, which determine the path, the energy, and the particle type, are often complex. The development and construction of these detectors, and the analysis of the data produced, are the province of the particle experimentalist. The strong interest in rare processes (such as W and Z production in the recent CERN experiments) and the need to characterize events completely have led to the development of detectors sensitive to almost the total solid angle. At all accelerators but particularly at particle colliders it is essential to provide detectors capable of making the fullest use of the particle beams. A wide range of detectors exists including the small but often sophisticated instruments designed for fixed-target work, the large detectors used for recording rare events such as the interactions of neutrinos or the decays of nucleons, and the large collider detectors that provide almost complete angular coverage and characterization of the interactions occurring in these machines. This latter class of collider detectors is among the most costly and demanding. Their technological problems and solutions are, in large part, shared with the other classes of

132

detector. The following discussion, for simplicity, will largely concentrate on the development, construction, and needs of this collider detector class. However, later in this chapter in the section on Detectors in Fixed-Target Experiments we present some highlights of the detectors used in these experiments. And in the final section of this chapter, we discuss experiments that do not use accelerators.

As an introduction to large detectors, we present the Mark I detector first used at SPEAR in 1972. It was the first electronic detector for a particle collider with close to full angular coverage and with a magnetic field to provide momentum and energy measurements for charged particles. A view of the Mark I is shown in Figure 6.1, and a reconstruction of a ψ (psi) event is shown in Figure 6.2. This detector was sophisticated for its era, allowing the discovery of the ψ, the tau lepton, and charmed particles.

A bare decade later comparably important results have begun to flow from the detectors at the CERN proton-antiproton collider with the discovery of the Z and W particles. Again these discoveries and the realization of the potential of the accelerator are only made possible by the sophistication of the detectors. The enormous advances in detector technology over the last decade are best illustrated by contrasting the view of a ψ event in the Mark I detector with the enormously more detailed picture from the UA1 detector at CERN of an event with a Z^0, shown in Figure 6.3. The actual configuration of the immense 5000-ton UA1 detector is shown in Figure 6.4.

Fixed-target detectors usually are more selective than the large collider detectors. The neutrino detectors are vast instrumented targets, usually incorporating magnetic analysis of produced muons and calorimetric measurement of produced hadrons. The target of a Fermilab neutrino detector, shown in Figure 6.5, has an instrumented mass of 690 tons, consisting of 20-cm iron slabs interleaved with drift chambers, followed by 420 tons of momentum-analyzing toroidal magnets. Such large masses are required to achieve a reasonable interaction rate from the weakly interacting neutrinos. In fixed-target experiments, the kinematics of high-energy relativistic collisions results in most of the final-state particles from an interaction being thrown forward into a relatively narrow cone. Consequently, fixed-target experiments generally appear as a linear sequence of detector elements downstream from the target, as Figure 6.5 clearly illustrates. We return to fixed-target detectors in a later section.

The larger detectors constitute major facilities, with a lifetime of usage of typically more than 10 years. They cost in the range of $10 million to $50 million and compete for resources with even the large

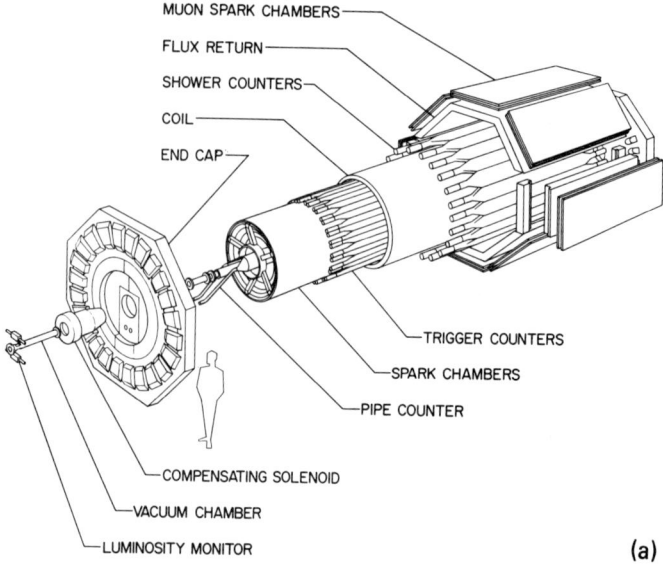

MUON SPARK CHAMBERS
FLUX RETURN
SHOWER COUNTERS
COIL
END CAP

TRIGGER COUNTERS
SPARK CHAMBERS
PIPE COUNTER

COMPENSATING SOLENOID
VACUUM CHAMBER
LUMINOSITY MONITOR

(a)

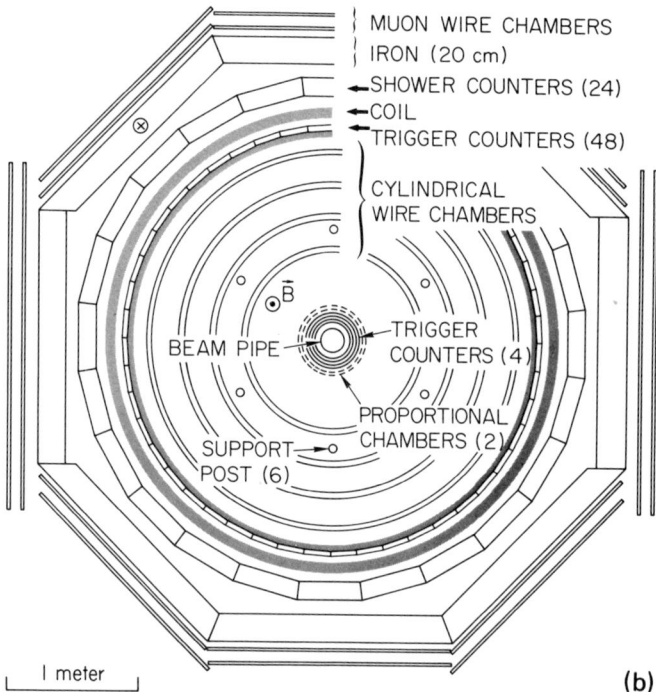

MUON WIRE CHAMBERS
IRON (20 cm)
SHOWER COUNTERS (24)
COIL
TRIGGER COUNTERS (48)

CYLINDRICAL
WIRE CHAMBERS

$\vec{B}$

BEAM PIPE

TRIGGER
COUNTERS (4)

PROPORTIONAL
CHAMBERS (2)

SUPPORT
POST (6)

1 meter

(b)

FIGURE 6.1   (a) The first general-purpose particle detector built for use at a particle collider was the Mark I detector shown here. (b) Cross-sectional view of the Mark I detector.

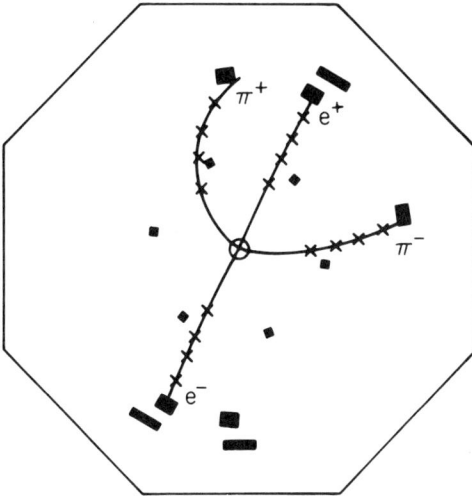

FIGURE 6.2   A computer reconstruction of an event found by the Mark I detector in which a ψ' particle decays to a ψ particle plus two pions; the ψ then decays to two electrons. The ψ' particle was in the small circle at the center, and the four lines coming out indicate the paths of the four particles produced in the decay. The lower-energy pion tracks are curved more strongly by the magnetic field. This particular picture became well known because the four paths also happened to form the Greek letter ψ.

parent accelerators. Unlike the early pioneering experiments of particle physics, a modern experiment may well require the simultaneous collaboration of several hundred physicists from 20 or more institutions. These major facilities require resources comparable with those used in the construction of the parent accelerator.

DETECTOR REQUIREMENTS AND PHYSICAL PRINCIPLES OF DETECTION

A detector system should be able to measure, as completely as possible, all the characteristics of the produced particles within an event. This implies that the detector should function over the largest possible angular range, measure with the best attainable precision, be provided with instrumentation to identify particle characteristics, and simultaneously provide a wide range of cross checks to protect against measurement artifacts. Additionally for use with hadron colliders, detectors must be able to extract interesting classes of physics events from backgrounds of events perhaps a hundred million times more frequent. These challenges must be met while simultaneously keeping

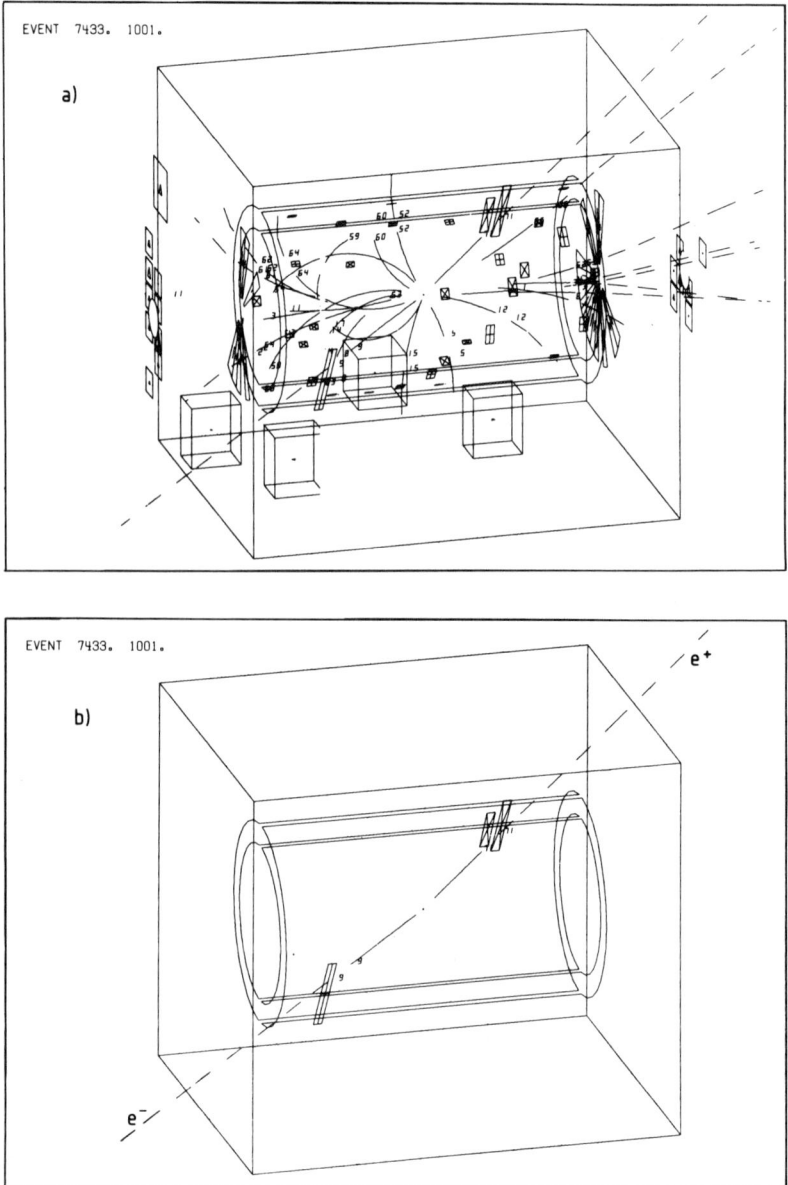

FIGURE 6.3   A computer reconstruction of an event produced at the CERN proton-antiproton collider; this event includes the production of a Z^0 particle. In a) the tracks of all the particles produced in the event are shown. In b) the two tracks of the electron and positron produced in the event are shown by themselves; this electron-positron pair comes from the decay of the Z^0.

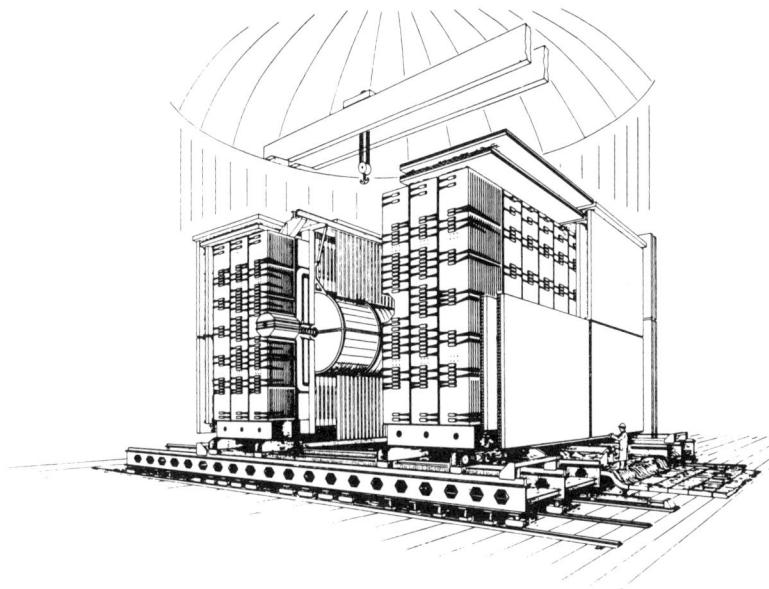

FIGURE 6.4 The UA1 detector at CERN, which produced the event shown in Figure 6.3. Note the enormous size of the detector compared with that of the person standing at its lower right side.

the combined costs of construction, operation, and data-reduction within reasonable bounds.

The basic physics underlying detector operation can be summarized as follows:

- Charged particles lose energy by ionization processes and leave a track or trail of ionized atoms and electrons as they pass through gasses, liquids, or solids. A wide range of techniques serves to measure the position and magnitude of these ionization trails. The magnitude of this energy loss per unit length is a measure of particle velocities.
- Velocities of particles may be determined from the time interval required to pass between two points. In general this technique differentiates velocities only for relatively small particle energies.
- In the presence of a magnetic field, charged particles are deflected into curved orbits. Measurement of this curvature permits the momenta of these tracks to be determined. The particle's energy can be calculated from its momentum if the mass is known.
- Characteristic, but weak, radiation is coherently emitted by particles passing through material—Cerenkov radiation; or by particles as

138

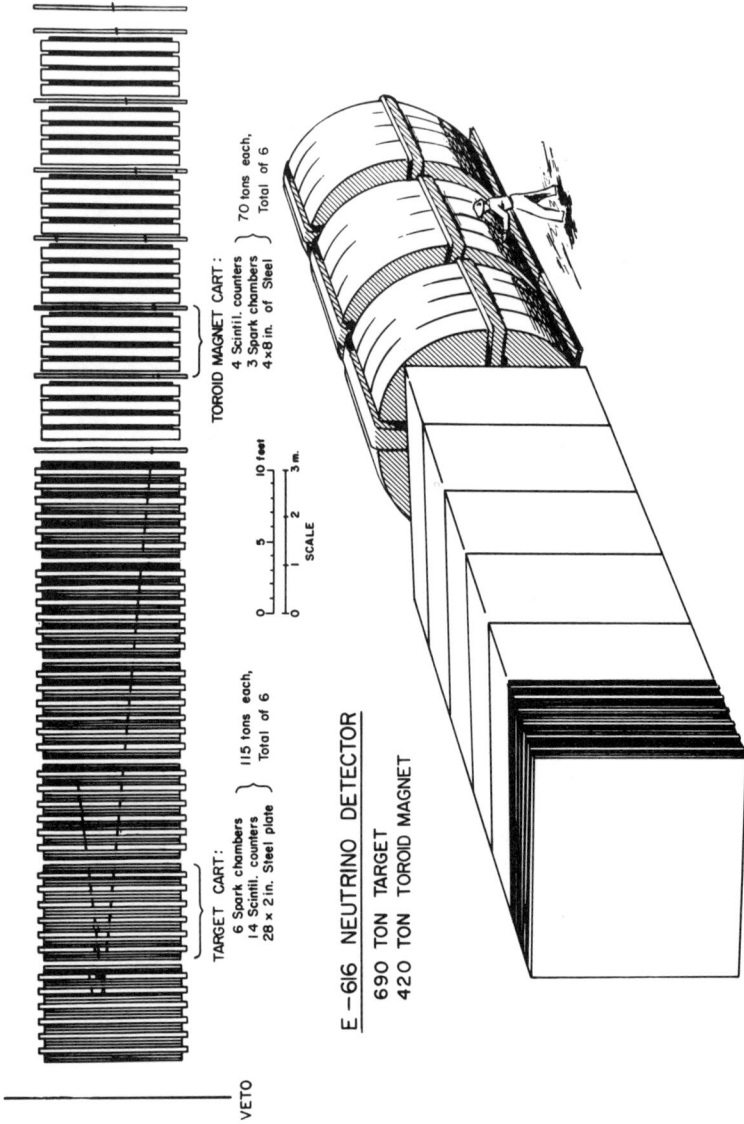

TARGET CART :
6 Spark chambers
14 Scintil. counters   } 115 tons each,
28 × 2 in. Steel plate   Total of 6

TOROID MAGNET CART :
4 Scintil. counters
3 Spark chambers   } 70 tons each,
4 × 8 in. of Steel   Total of 6

VETO

SCALE
0   5   10 feet
0   1   2   3 m.

E – 616 NEUTRINO DETECTOR
690 TON TARGET
420 TON TOROID MAGNET

FIGURE 6.5 A detector used for studying neutrino interactions at Fermilab. Since neutrinos interact rarely, a very large and dense target, 690 tons of iron, must be used.

they cross an interface between different materials—transition radiation; or as they pass through magnetic fields—synchrotron radiation. The intensity and characteristics of these radiations can serve as the basis of a velocity measurement.

• When electrons or photons pass through matter they produce characteristic electromagnetic cascades of secondary radiation, which in turn leave an intense core of ionized atoms. The energy of the original electron or photon is ultimately completely converted into ionization by these processes. Measurement of this converted energy in a sufficiently thick block of material determines the total incident electromagnetic energy and constitutes a calorimetric shower-energy measurement. A detector constructed to make use of this property is an electromagnetic calorimeter.

• Hadrons, such as protons and mesons, interact strongly as they pass through matter, producing secondary hadrons. This again results in the production of intense cores of ionized atoms. The total energy of the incident particles can be measured from the total energy deposited in the form of ionization. Hadronic cascades can be differentiated from the electromagnetic cascades that develop in much thinner layers of material. The technique of measurement is known as hadron calorimetry. Typically a hadron calorimeter might require a thickness of 3 to 4 feet of instrumented steel with a total weight of hundreds of tons.

• Energetic muons are uniquely characterized by the property that, although charged, they penetrate large thicknesses of material and emerge with a relatively small change of energy at the outside of a detector.

DETECTORS FOR COLLIDER EXPERIMENTS

Modern detectors typically make use of all or many of the above properties to characterize detected events. The characteristics of these detectors have many features in common and share similar design architectures. The detectors for collider experiments are based on a series of concentric shells or layers, one behind the other, each of which is devoted to some particular aspect or aspects of the detection process. The initial detector layers are used to characterize the charged-particle component and are designed to be nondestructive, i.e., to contain little material so that charged particles will not interact or degrade in energy and thus will maintain their identity while traversing the layers. The outer detector layers deliberately use large amounts of material in order to materialize the neutral particles and to convert the energy carried by the particles into detectable ionization; such a device is known as a calorimeter.

The outer layer of the calorimeter or an additional detection layer is frequently used to detect muons. As mentioned earlier, energetic muons are usually the only particles that can reach this outermost layer.

Inevitably, as the layered levels of detection systems are built up, a detector will become large and correspondingly complex and expensive. A prime objective of detector development is, therefore, to keep detection systems as compact as possible and to combine detection roles whenever possible.

Additional demands are imposed on detector systems associated with hadron colliders by the high ambient radiation levels at the detector and by the fact that events of interest may be separated only by short times from uninteresting background events.

Summarized below are the elements or layers constituting a typical modern detector system and some of the ongoing research and development aimed at maximizing present or future detector capabilities.

Close-in Detection: Vertex Detectors

A fraction of the particles emerging from a collision point decay in flight at distances as close as 0.001 cm. Such decays provide characteristic signatures as to the nature of the decaying particles. Therefore use can be made of charged-particle detectors with high spatial resolution that are placed as close to the interaction point as possible. The first such vertex detector for collider work was recently constructed to operate with the Mark II detector at the PEP collider at the Stanford Linear Accelerator Center. Several such detectors are now being constructed or are operating at electron-positron and proton-antiproton colliders.

The first generation of such vertex detectors was based on conventional track-detection methods for charged particles, using multiwire drift chambers or time-projection chambers (TPC). (The principles of operation of these track detectors are described below.) The chambers are typically only about 10 cm in radius, are fabricated with fine subdivisions to provide separation between adjoining tracks that might otherwise overlap, are aligned with great precision (about 0.002- to 0.005-cm tolerance), and ultimately are likely to be operated under high pressures to provide sharp internal localization of the trail of ionization left by the charged particles. The limits of precision for such detection are at present about 0.01 cm; this should eventually improve by a factor of 2 or 3.

The second-generation vertex detectors now under development are based on modern silicon semiconductor technology. This technology has already been used successfully on a small scale in experiments designed to measure decays of short-lived particles, which decay close to the parent event. In these experiments the target is constructed of microstrips, and a detailed history of events occurring within these active targets can be recorded. Preliminary tests have been made and have established the feasibility of the proposed vertex detectors. Full-scale detectors should come into operation during the next 5 years. Three approaches are being tried: (1) the use of packages of long thin silicon strips with widths of about 0.002 cm, (2) the use of a mosaic of semiconductor squares as currently exist in the charge-coupled devices (CCD) that are used as a basis for image detection in astronomical and other applications, and (3) a more speculative idea that involves drifting the ionization over a relatively long distance within the silicon.

The handling, precision alignment, and electronic readout of such miniaturized devices present fascinating but soluble problems. With reasonable confidence this second generation of detectors should provide precisions an order of magnitude better than those currently obtainable.

Charged-Particle Tracking Chambers

In a typical collider detector, beyond the vertex detector are charged-particle tracking chambers. These chambers serve to measure the directions and curvatures of the paths of the individual particles. These paths are called tracks. The principle of operation of a multiwire drift chamber is shown in Figure 6.6. The first such devices were coarse and measured only a few tracks. Modern devices are fine grained in subdivision and may provide over a hundred measured points to a track. The resulting image is almost of photographic quality and is reminiscent, even though produced at electronic speeds, of the superb track detection provided in bubble chambers. With some additional effort and with certain possible compromises these detectors can also be used to measure the ionization of the produced tracks.

An elegant variant of this detection method is to remove the fine grid of wires and to drift the ionization with a collection electric field to the end caps, where the arrival positions of the ionization are measured and also the arrival times and the degree of ionization. The arrival

```
+HV ————————————————

-HV ————————————————

+HV ————————————————    ⎰Signal Appears
                         ⎱on This Wire
            ●
-HV ————————————————

+HV ————————————————

-HV ————————————————

+HV ————————————————
```

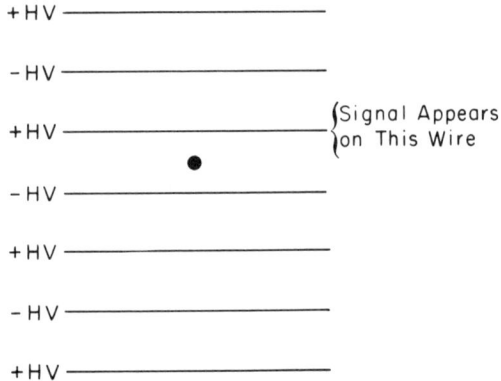

FIGURE 6.6 Most detectors use drift chambers or devices derived from drift chambers. In a drift chamber, slender wires are strung parallel to each other in a volume of gas. When a charged particle passes through the gas it leaves a track of ionized gas molecules and electrons. Electrons are attracted by the electrical voltage on the wires, and when they reach the wire they send an electrical pulse down the wire. Those pulses are collected and amplified electronically and recorded on magnetic tape. The position of the track is given roughly by knowing the wire that gave the signal. But a more accurate position is obtained knowing the time the electrons took to drift to the wire, hence the name drift chamber.

times provide a measure of the depth at which the particle was produced, because the ionization drifts under the influence of the collection fields with a fixed velocity. This system, known as the time-projection chamber (TPC), has been implemented and is currently used in the TPC detector at PEP.

Charged-particle tracking detectors are often immersed in a magnetic field in order to make it possible, from measurements of the track curvatures, to determine the signs of the charges and the momenta of the particles. Magnetic fields in the range of several kilogauss to several tens of kilogauss are used. The larger the field, the more the tracks curve, and the easier it is to measure the track momentum. To provide the highest possible magnetic fields, it is desirable to use superconducting coils to carry the required large currents. Although a number of these coils have been constructed and successfully operated, the technology of fabrication is demanding, and further research is desirable.

Identification of Particle Types

Next in sequence beyond the tracking chamber there may be a detector layer that is used to identify the nature of the charged particles (whether they are electrons, protons, pions, kaons, or muons). This region is still required to be nondestructive and thus to contain little material. A number of identification methods have been tested on small-scale devices and are under development for the next generation of detectors. These include the following:

• Particle identification using Cerenkov radiation. This technique makes use of the property that particles produce light at an angle that depends on the velocity with which they are traveling through transparent matter. This radiation can be focused to a ring image whose radius directly measures the particle velocity. Counters photoconvert this ultraviolet radiation into ionization, which is then detected with proportional-counting techniques. This technique has been demonstrated successfully on a moderate scale but still requires considerable development to make it viable for large-scale detection.

• Electron identification via the phenomena of transition radiation or, alternatively, via the emission of synchrotron radiation in the magnetic fields traversed.

Calorimetric Detection and Energy Measurement

The detection systems described so far do not serve to detect neutral particles, such as photons and neutrons, nor would they permit a precise measurement of the total energy in an event. The final layers of a collider detector therefore comprise thick, highly instrumented blocks of material—calorimeters—in which electromagnetic and hadronic particles cascade and convert their energy into ionization. These final instrumented blocks are placed at large radii, away from the point at which the beams interact, to leave sufficient space in which to insert the nondestructive low-density systems. The large radii of these blocks, coupled with the requirement of substantial thicknesses, result in calorimeters that are massive objects.

The calorimeters used in collider experiments divide naturally into a front region where most of the electromagnetic cascade energy is deposited—the electromagnetic calorimeter—and the back region where most of the hadronic cascade energy is deposited—the hadronic calorimeter. Maximum precision is obtained when the calorimeters are constructed of materials, such as sodium iodide, in which the total deposited energy is detectable. However, such active calorimeters for collider detectors would, with present technology, be prohibitively expensive. It is still very desirable to construct at least the electromagnetic calorimeter out of active material. Some of the materials under development, or in use, for electromagnetic calorimetry are heavy glasses, bismuth germanate crystals, and barium fluoride crystals. These systems are still costly, and their use at present is only made possible by leaving less space for the low-density systems in order to minimize the material requirements. Further developments in the production of comparatively low-cost materials for use in electromagnetic calorimeters would be useful.

Even with substantial advances, however, most electromagnetic calorimeters and all hadronic calorimeters are likely to continue to rely on the introduction of many active sampling layers interspersed throughout the large passive calorimeter block in order to measure the ionization produced. Another design goal that is important but hard to realize is a finely divided calorimeter that is able to provide precise locations of the deposited energy. This requirement follows from the fact that the particles emitted from events in high-energy colliders are tightly bunched into jets. Important information can be extracted from the angular spread and characteristics of the energy deposited by these jets of particles. This fine division or segmentation typically may require the recording of information from many hundreds or thousands of electronic channels reading out the information from the individual cells.

DETECTORS IN FIXED-TARGET EXPERIMENTS

The detectors used in fixed-target experiments vary tremendously in design, in size, and in complexity. Since we cannot do justice here to the range and variety of these detectors, we will only give several examples. The examples can be brief because the components of these detectors are in general just the same elements that we have been describing. The major exception is the bubble chamber, which is discussed at the end of this section.

Small or Simple Fixed-Target Experiments

Fixed-target experiments can sometimes be carried out with small or simple particle-detection equipment. This is often the case when the physicist is studying a simple reaction of elementary particles or studying one particular property of a particle. Two examples are given

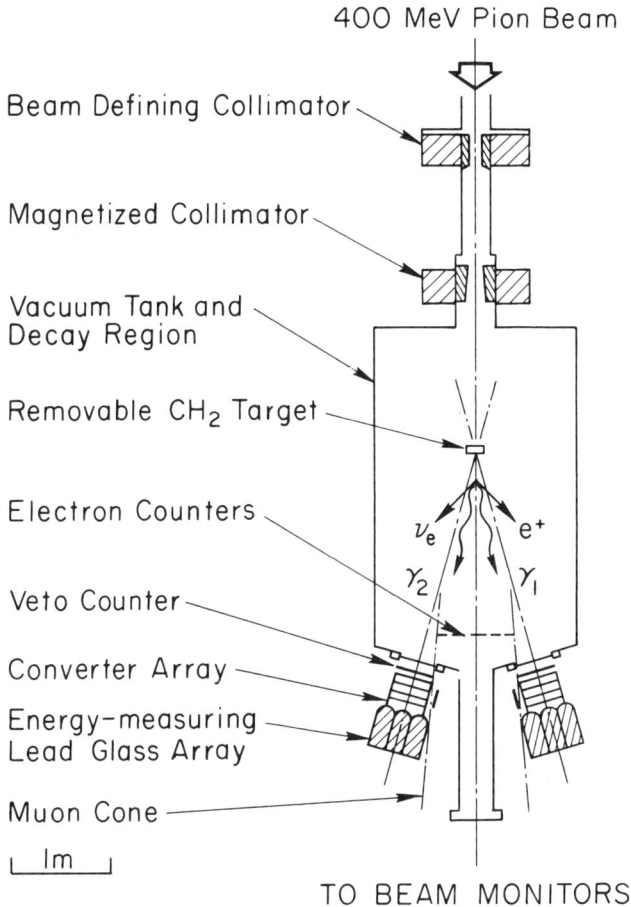

FIGURE 6.7 This apparatus, used at the Los Alamos LAMPF 800-McV proton linear accelerator, studies the decay of a charged pion to a neutral pion plus a positron (e^+) plus a neutrino (ν_e). The neutral pion decays to two photons (γ_1 and γ_2).

in Figures 6.7 and 6.8. The first example is an apparatus used at the Los Alamos LAMPF 800-MeV proton linear accelerator. The proton beam is used to produce a charged pion beam, and the pion decay process

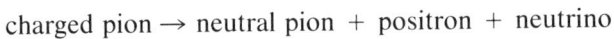

charged pion → neutral pion + positron + neutrino

is studied. The apparatus is relatively small and primarily uses two electromagnetic calorimeters, yet this is a fundamental measurement.

Figure 6.8 shows a neutrino detector used at the AGS proton accelerator at Brookhaven to test the stability of muon neutrinos. The

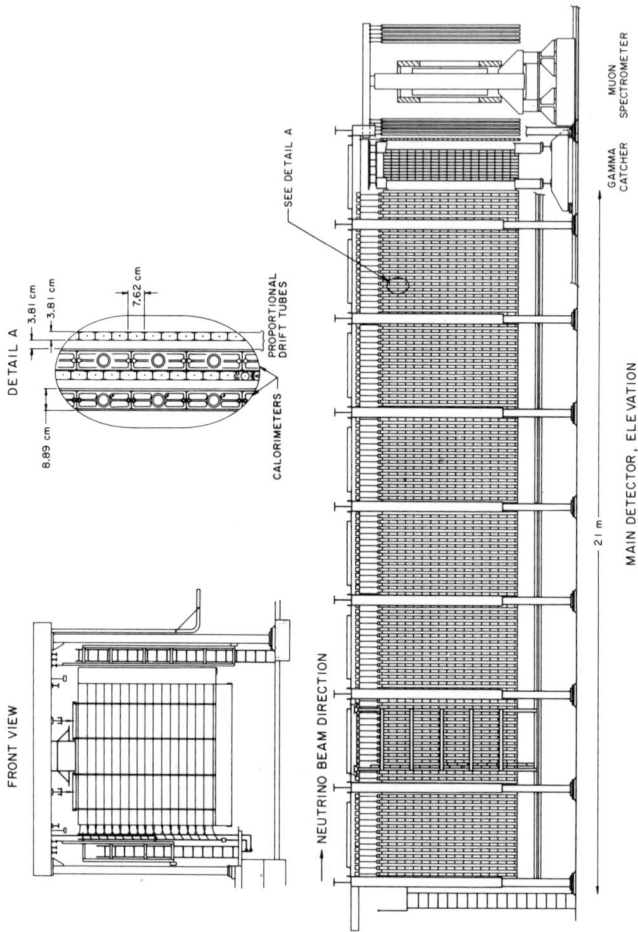

FIGURE 6.8 A neutrino detector at the AGS proton accelerator at Brookhaven. This 175-ton detector is composed of 112 identical modules, each containing a vertical array of scintillator cells and measuring planes of proportional drift tubes, followed by a shower counter and a magnetic spectrometer. It is used to measure the elastic scattering of neutrinos and antineutrinos on electrons and protons and to search for neutrino oscillations.

muon neutrinos, produced by the primary proton beam, are allowed to travel a long distance before striking the detector. The detector is of moderately large size but simple construction; its function is to detect neutrinos and to distinguish between muon neutrinos and electron neutrinos.

Large or Complex Fixed-Target Experiments

A major fixed-target facility that demonstrates many instrumentation techniques is the Fermilab Tagged Photon Spectrometer, shown in Figure 6.9. It is intended primarily for studying the production of charmed particles. A photon beam produced by the primary proton beam strikes a liquid hydrogen target. Recoiling protons are identified by the recoil detector, and the produced particles are analyzed in the forward spectrometer. The spectrometer has magnetic analysis to measure charged-particle momenta; Cerenkov counters to identify pions, kaons, and protons; electromagnetic calorimetry to measure the energy of neutral hadrons; and finally a set of scintillation counters behind an iron filter to detect penetrating muons. This sequence of analysis steps is the same as that used in most collider detectors, but the target is not surrounded by all the components of the detector as it is in a collider detector.

Figure 6.10 shows a rather complex detector that combines modern detector elements with the old nuclear emulsion technique. A nuclear emulsion is a thick photographic emulsion that when developed shows the paths of charged particles that have passed through it. This detector was used at Fermilab to measure the lifetimes of charmed mesons. The emulsion gave precise pictures of how the mesons decayed close to their production point.

Bubble Chamber

The bubble chamber, invented in the 1950s, was for many years the workhorse of elementary-particle physics experiments. A bubble chamber uses a superheated liquid, such as liquid hydrogen, neon, or Freon. Charged particles passing through this liquid leave tracks of tiny bubbles, which are photographed. The bubble chamber has gradually been replaced in most experimental applications by electronic detectors. The latter are more versatile, often give more information about the products of the collision, and usually provide that information in a form that can be directly used in computers. Nevertheless, the large volume and precise track information provided by bubble chambers

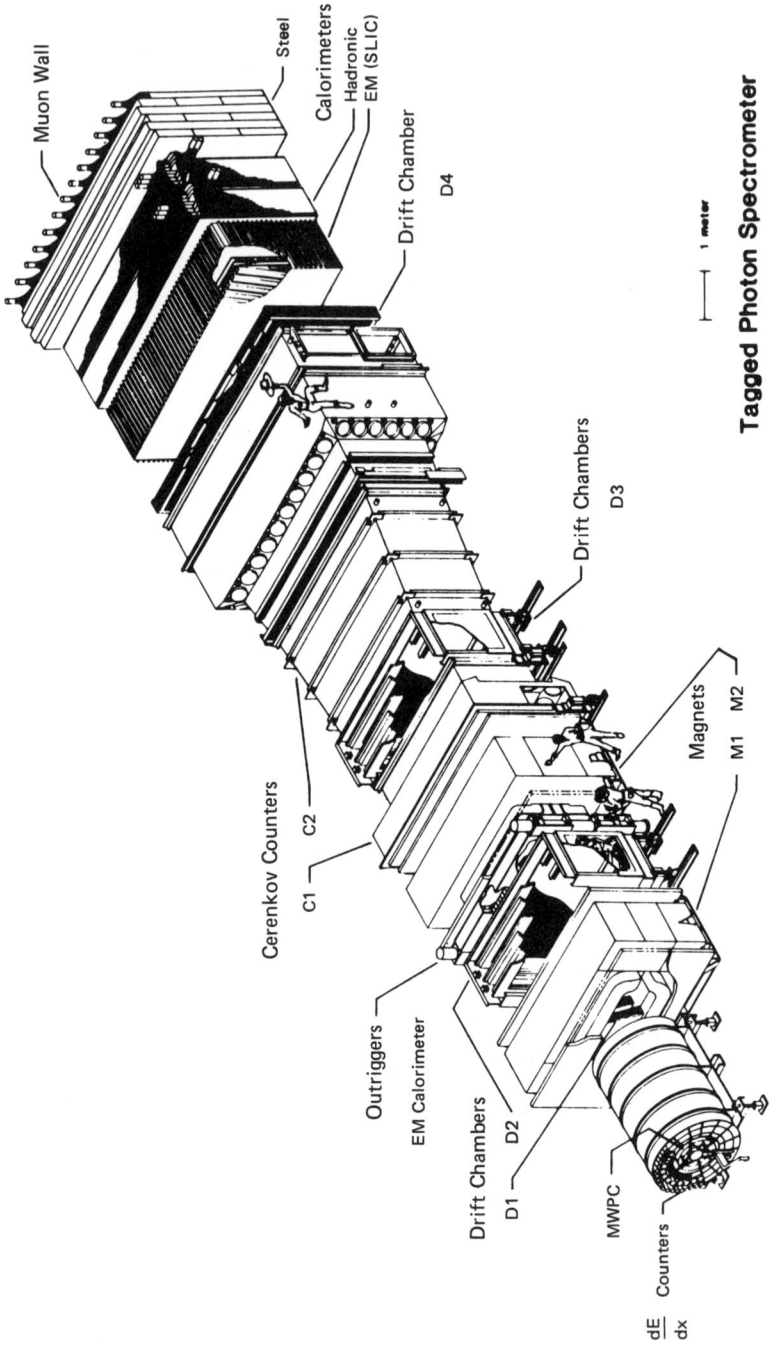

Muon Wall

Steel

Calorimeters
  Hadronic
  EM (SLIC)

Drift Chamber
D4

Cerenkov Counters
C1   C2

Drift Chambers
D3

Outriggers

EM Calorimeter

Drift Chambers
D1   D2

MWPC

Counters

$\frac{dE}{dx}$

Recoil Detector

Magnets
M1   M2

1 meter

Tagged Photon Spectrometer

FIGURE 6.9   This detector facility at Fermilab is used to study the production of charmed particles produced when photons interact with hadrons.

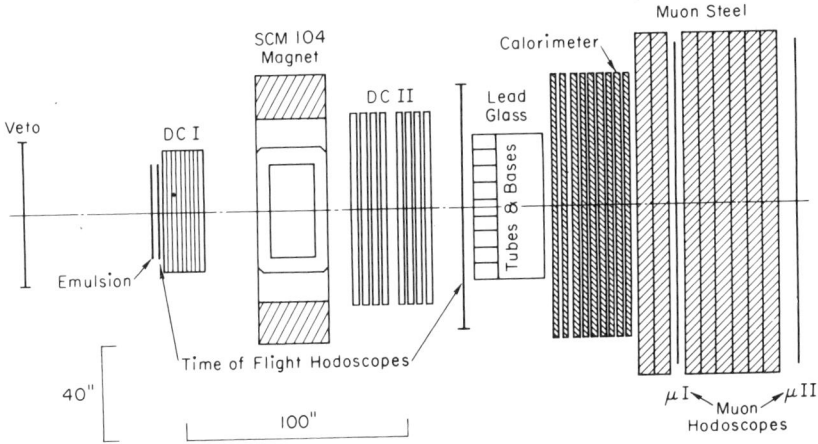

FIGURE 6.10 This fixed-target particle detector combines the old technique of the nuclear emulsion with new techniques of chambers. Used to measure the lifetime of charmed particles, the emulsions give a precise location for where the charmed particle decayed.

still makes them most suitable for certain types of experiments. Chief among these are the study of neutrino interactions and the study of particles with short lifetimes. Recent improvements in bubble-chamber technology include high repetition rates, precise track measurements, and holographic photography.

DATA REDUCTION AND COMPUTERS

Experiments in high-energy physics characteristically have produced great quantities of data, whose reduction and physical interpretation have made up a significant component of the experimental effort. In recent years, apart from the sheer increase in scale of these experiments, data reduction has evolved to become more integrated with and intrinsic to an experiment's operation. A particular example is Monte Carlo computer simulation, which has become an important means of experimental calibration. New detector capabilities have made this evolution both possible and necessary. Very precise time resolution (billionths of a second, in some cases) has become possible over large detection volumes, allowing selective recording of particular event classes that make up only a tiny fraction of the total rate. In turn, these fast detectors can supply information in a form that can be rapidly digitized and processed to provide the criteria for real-time event selection.

Advanced systems for the reduction of high-energy data have unified the traditionally separate functions of trigger decision making, data logging, experimental control, and off-line event analysis. The trigger is critical to the success of many experiments because collisions may occur at a rate exceeding 10^6 per second. In order to select those interactions that are of particular interest in an experiment, a trigger is used. The trigger is a fast electronic decision made to record the data from a particular event on the basis of the signals received from the detector.

The triggering decision itself may be based on detector input that no single computer could process fast enough. Fortunately, the repetitive nature of these calculations can be adapted to the use of fast but relatively primitive processors working in parallel. These are the first steps in what traditionally would have been termed off-line analysis. Without the results of these and other computations, the operation of advanced detectors cannot be monitored or controlled. Even the logging of data onto tape may use many levels of the data-acquisition system. For example, a single trigger may contain 100,000 characters of data (the equivalent of a 30-single-spaced-page report), which must be assembled from widely distributed local memories into the image of one event. One experiment in one year can accumulate data amounting to 10 million such reports. Because of the enormous software development required, programs for real-time and off-line event analysis increasingly must share common subroutines and other features. The distinction between real-time and off-line analysis is further blurred by the scale of processing power that must be dedicated to a single experiment, in notable cases reaching a level comparable with that of a major computer.

The trend toward large-solid-angle general-purpose detectors at the colliding-beam facilities has been noted. As the collider energy rises, the events become more complex, and the amount of computer time required to analyze these events becomes large. For example, a Z^0 decay into hadrons has an average of about 20 charged particles and 22 neutral particles, and the off-line analysis time for such events in the detectors proposed for the SLC or LEP is of the order of 100 seconds of central processor unit (CPU) time for moderate-size computers. Millions of such events per year may need to be processed. As another example, it has been estimated that a dedicated capacity equivalent to tens of moderate-size computers will be required to process the interesting events from the 2-TeV proton-antiproton collider at Fermilab. It is expected that the computer requirements for the high-luminosity SSC machine will greatly exceed the present level, as event rates, complexity, and detector sophistication will all increase.

These requests for computer power are marginally met by current commercially available computers. The supercomputers tend to be machines optimized for vector or array calculations typically encountered in solving large sets of partial differential equations, such as in weather prediction. These machines are not well suited to the large input-output (IO) requirements of high-energy physics nor to the large address-space requirements of the detector analysis codes. Some manufacturers have addressed the IO problems and have reasonable CPU power but are relatively weak in modern software tools and in the system architecture to support them. These are important issues in high-energy physics, where the data production codes are being constantly improved, and data analysis involves significant amounts of programming, almost all of which is done by the physicist. Tools that improve the efficiency of the physicist are clearly valuable. Finally, the manufacturers of superminicomputers who have advanced the software state of the art do not produce machines of sufficient CPU power to analyze the data from a modern collider detector.

The present situation is sufficiently serious to motivate noncommercial attempts to provide adequate computing power. Most of these attempts are based on the relatively large fraction of CPU to IO activity that characterizes detector event analysis, so that many relatively simple, laboratory-designed processors can work on events in parallel, while being controlled by one commercial host computer with extensive IO facilities. Examples are the emulator developments at SLAC and CERN and the multiple-microprocessor project at Fermilab. Related projects involve even more specialized processors designed for lattice gauge theory calculations or accelerator ray tracing. It is, of course, possible that commercial developments will prove adequate in the next few generations of machines, but the present situation is murky. Computing represents a significant expense at the national laboratories and universities in terms of actual hardware, support personnel, and physicist involvement. Even with the rapid decline in the unit costs of computing (crudely a reduction by a factor of 2 every 3 years), the overall costs of computing increase.

Thus computers of both large and medium size are now necessary parts of almost all elementary-particle physics experiments. Not only are they needed to reduce and study the data, but they are also used to monitor and control the experimental apparatus. The extensive use of computers has stimulated advances in some types of computer technology and systems. This is because physicists have been willing to work with computers that were still in an early stage of production, interacting closely with the computer manufacturers.

FACILITIES AND DETECTORS FOR EXPERIMENTS NOT USING ACCELERATORS

Most particle-physics experiments use accelerators, but some do not. In this section we sketch some of the ways in which elementary particles are studied without using accelerators.

Atomic, Optical, Electronic, and Cryogenic Experiments

Elementary-particle physicists are concerned with precise measurements of the properties of the more stable particles, such as electrons, positrons, muons, protons, and neutrons. For example, the electric charge of the proton is the same magnitude as the electric charge of the electron according to the most precise measurements that can be made. This is one of the reasons why we believe that there is a connection between the leptons (the electron is a lepton) and the quarks (the proton is composed of quarks). Such precise measurements are carried out using the methods of atomic, optical, electronic, or cryogenic physics. These include measurements of the electron g-factor, of the positronium Lamb shift, and of parity violation in atomic systems.

These methods are also used to search for new types of particles in matter. Two sorts of searches have particularly intrigued particle physicists. One is the search for free quarks as contrasted with the quarks that are bound together inside protons and neutrons. The other is the search for magnetic monopoles, that is, for an isolated magnetic pole. All known particles that have magnetic properties have two magnetic poles, one north and one south, of equal size. None of these searches has produced generally accepted evidence for free quarks or monopoles. But more definitive searches are under way.

Experiments Using Radioactive Material or Reactors

There are experiments using radioactive material or reactors that are important in both elementary-particle physics and nuclear physics. An outstanding example is the study of the radioactive beta decay of tritium. An electron and a neutrino are produced in this decay, and the mass of this neutrino can be measured. The mass of the neutrino is a pressing question: is it exactly zero or, as indicated in a recent tritium-decay measurement in the Soviet Union, is the mass nonzero? Another example involves different forms of beta decay that have been proposed, such as the production of two electrons but no neutrino. This would violate our present theory of the weak force, and thus it is important to ascertain if such a decay exists.

Reactors produce electron neutrinos, which are being used to study the stability of these particles. The most recent experiments find no confirmed evidence for neutrino instability. Incidently, evidence for neutrino instability is also being sought with solar neutrinos, as described below, as well as with neutrinos from accelerators. This illustrates how the exploration of an area in particle physics spreads over all the experimental techniques. Reactor experiments have also set important upper limits on the neutron's electric dipole moment.

Experiments Using Cosmic Rays

Cosmic-ray physics is concerned with three areas: the origin of cosmic rays, the use of cosmic rays as probes of extraterrestrial phenomena, and the use of cosmic rays to study elementary particles. It is the third area that concerns us here.

Earlier in this century, cosmic rays were the only source of very-high-energy particles; hence substantial discoveries in particle physics were often made with cosmic rays. Prominent examples are the discoveries of the positron, the muon, and some of the strange hadrons. However, accelerator experiments have gradually displaced cosmic-ray experiments. Cosmic-ray experiments at present can only contribute to elementary-particle physics at extremely high energies, but unfortunately the flux is then small. This is shown in Figure 6.11.

The most ambitious facility for the study of the highest-energy cosmic rays in the United States is the University of Utah's Fly's Eye detector. Here a matrix of about 1000 phototubes in two clusters separated by 3.3 km observes, with good spatial and time resolution, the atmospheric scintillation light from cosmic-ray air showers. These air showers come from the interactions of very-high-energy cosmic-ray protons (energies above 10^8 GeV) with nuclei in the upper atmosphere.

Because the cosmic-ray flux is so low at the highest energies (less than one per year per km^2 above 10^{10} GeV), it seems impractical to study these events in any way except through the study of air showers. Even an ambitious space station would not be able to support a detector sufficient to address this energy regime. By utilizing the atmosphere as a target, the Fly's Eye technique can explore a very large area, of the order of 100 km^2 or greater. If the detector were high

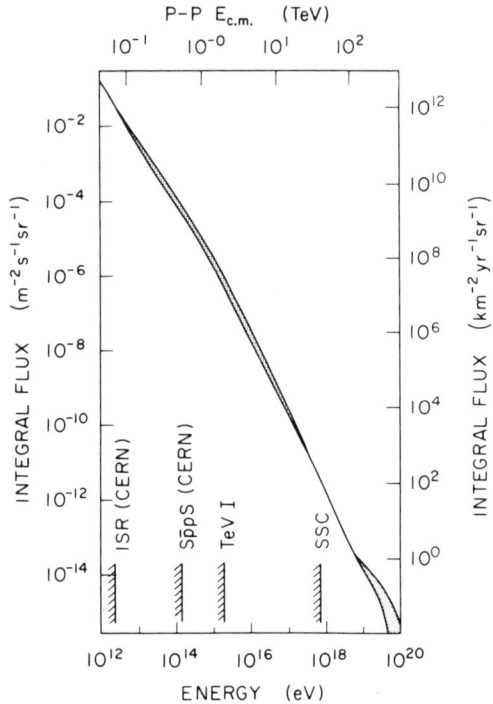

FIGURE 6.11   The integral cosmic-ray flux, the number of primary cosmic-ray nuclei of energy greater than E, plotted versus E expressed in electron volts (eV). The vertical scale is expressed as particles per square meter per second per steradian (left) and as particles per square kilometer per year per steradian (right). The energy scale is also given in terms of the equivalent proton-proton center of mass energy, $\sqrt{s}$. The shading represents the experimental uncertainty of the flux determinations.

in the atmosphere, the details of the first interaction would be more accessible. Again, this approach is the *only* access to particle physics at energies greater than even a 40-TeV proton-proton collider can provide, which is equivalent to about 10^9 GeV (10^{18} eV) for cosmic-ray interactions.

Turning to another example, current unified theories predict that there should be massive magnetic monopoles with a rest mass of about 10^{16} times the proton mass and further that these should have been produced in the early universe and still be present among cosmic rays. An experiment in early 1982 reported evidence for the passage of one such monopole through a superconducting coil of a few cm^2 area. A large number of other experiments, using larger superconducting coils and also searching for slow, lightly ionizing particles, have subsequently failed to see evidence for monopoles. Larger monopole detectors are now being built.

Another volume in this survey describes cosmic-ray experiments in more detail, with particular emphasis on their impact on questions in astrophysics and cosmology.

The Solar Neutrino Experiment

A major nonaccelerator facility with ramifications in particle physics, nuclear physics, and astronomy is a Brookhaven National Laboratory detector for solar neutrinos. This detector is seeking evidence for inverse beta decay produced by neutrinos from the Sun. This process, whereby a neutrino produces a transition of a chlorine nucleus to an argon nucleus, would be a clean signature for solar neutrinos. After some years of operation and a multitude of careful checks, the observed rate of argon production is only about one third to one fifth of the predicted rate. Either our nuclear physics and astrophysics understanding of the solar furnace in which hydrogen is burned in thermonuclear reactions is incorrect or some of the neutrinos decay before they reach the Earth (and neutrinos are unstable) or the experiment is wrong. Although this important problem remains unsolved, no new solar neutrino detectors are now being built. However, a new kind of solar neutrino detector using gallium has been proposed and designed in detail. Such a detector would be sensitive to the lower-energy neutrinos coming from the Sun. This is an advantage, since the number of these neutrinos predicted by theory is less dependent on a complete understanding of the conditions in the interior of the Sun.

Searches for the Decay of the Proton

Until the last decade the proton was regarded as absolutely stable; that is, it was assumed that the proton could not decay to any other particle. However, the realization that quarks, and hence the proton, are related to leptons has been growing. Therefore we have begun to consider the possibility that the proton could decay to either an electron or a muon (both leptons) plus other particles.

This possibility, made quantitative by grand unification theories, has led to a first-generation family of underground proton-decay experiments. As of the end of 1983, the earliest very large detector, built by the Irvine-Michigan-Brookhaven (IMB) group, had seen no evidence of proton decay in about 4000 tons of water. This implies that, depending on the mode of decay, the proton lifetime is probably greater than 10^{31} to 10^{32} years. This detector employs about 2000 photomultipliers mounted on the walls of an 8000-m^3 water tank. The signals are from Cerenkov radiation in the water.

Several large detectors in Europe, India, Japan, and the United States are or soon will be operating with only somewhat smaller masses. Some of these detectors use ionization for tracking and calorimetry, rather than the Cerenkov technique, and will therefore be more sensitive to some decay modes than the IMB experiment. Together with further IMB data these detectors should either identify proton decay or set lower limits to the proton decay lifetime of between 10^{32} and 10^{33} years for each of several expected decay modes.

SUMMARY AND FUTURE PROSPECTS

In the preceding discussion a number of trends in instrumentation and detection systems for high-energy physics have been identified. Jets signatures produced by quark interactions at high momentum transfers have led to great emphasis on finely segmented detection systems and to calorimetric detectors for the measurement of energy flow. New generations of quarks and leptons with lifetimes in the range of one trillionth of a second have stimulated new progress in the development of electronic systems for high-resolution vertex detection. Emphasis on rare processes has required virtually all major new detectors to be designed for efficient operation over nearly the full angular acceptance. For successful detection under conditions of high ambient rate, new levels of sophistication and integration of data-acquisition systems have become necessary.

Continued progress in the development of instrumentation and detection systems is the experimental high-energy physicist's greatest challenge: these systems provide both the *raison d'etre* for accelerator facilities and the means by which our theoretical understanding can be confronted by experiment. Support for the development of high-energy physics instrumentation should grow in breadth and intensity. Basic research in the development of new detectors for high-energy physics increasingly should be recognized and supported as a fundamental source of much progress in the field. Correspondingly, the experimental stations at our front-line accelerators should be used as effectively as available technology will reasonably permit.

7

Interactions with Other Areas of Physics and Technology

The purpose of research in elementary-particle physics is to investigate the basic nature of matter, energy, space, and time. In the course of this research elementary-particle physics interacts with other areas of physics. Subject matter, instruments, and theories from other areas all have a bearing on elementary-particle research. And elementary-particle research contributes data, theories, and apparatus to other parts of physics. In this chapter we briefly describe the interaction between elementary-particle physics and four other areas: cosmology and astrophysics, cosmic-ray physics, nuclear physics, and atomic physics. Those areas are the subjects of separate volumes in this survey; here we look only at their interaction with particle physics.

Elementary-particle physics interacts with technology in two ways. First, the technology that is invented and developed for use in particle physics subsequently finds use in other fields. The foremost example is the particle accelerator itself, some of whose applications are described below in the section on Other Applications of Accelerators. The second way that elementary-particle physics interacts with technology is that technology from outside particle physics is stimulated and developed during the design and construction of particle-physics accelerators and detectors. Prominent examples are superconducting magnets, described below in the section on Large-Scale Uses of Superconductivity, and integrated circuits and computers, described in the section on Support and Stimulation of New Technology.

COSMOLOGY AND ASTROPHYSICS

Recent years have witnessed a growing symbiotic relationship between elementary-particle physics, the science of the very small, and cosmology, the science of the very large. This interdependence has been fostered by the revolution in our understanding of particle physics on the one hand and on the other hand by the existence of and observational support for the big bang model of the origin of the universe. In that cosmology, the universe evolved from an original explosion of a dense, hot mixture of matter and energy. In those first moments of the universe elementary particles were created and destroyed at an enormous rate. Their creation and destruction occurred through interactions. For example, if two photons collide they can interact and be destroyed, but as a result of that interaction an electron and a positron can be created. In big bang cosmology the universe is expanding as this happens, the matter and energy cooling off and becoming less dense. Some particles no longer interact, hence they are not destroyed, and they remain in our present universe. Such particles are called relics.

This is the natural interpretation of the observed recession of distant objects in the universe, and it is supported by the observations of the 3-K microwave background radiation, generally interpreted as a relic from the recombination of electrons and nuclei to form electrically neutral matter. Recombination occurred when the universe was thousands of times more compressed than it is today, with a temperature of several thousands of kelvins. Such an increase of temperature at earlier times when the universe was denser receives further support from successful calculations of the astrophysical abundances of light elements through cosmological nucleosynthesis when the universe was a billion times smaller than it is today, with a temperature of several billion kelvins.

The successes mentioned above illustrate the principles enabling big bang cosmology and particle physics to be interrelated. Our present understanding of particle physics enables us to extrapolate the Hubble expansion of the universe backward to earlier times and higher temperatures and to calculate the abundances of other elementary-particle relics from the big bang. An example is provided by the question of how many kinds of neutrinos exist. This is a critical question because it is one way to find out the number of generations of leptons that exist. Nucleosynthesis calculations impose the most stringent available limit on the number of light neutrino species. They are restricted to three or at most four, which is below the best upper limits currently available from particle-physics experiments.

As another example, from present theories it can be calculated that stable neutrino masses must be less than 100 eV, or more than a few GeV, if these neutrino relics are not to decelerate or reverse the present expansion of the universe because of their mass. These constraints on neutrino masses are much more general than those obtained from laboratory experiments, and for the muon and tau neutrinos they are more stringent.

Another long-standing problem in cosmology has been the existence of dark matter; that is, the amount of visible matter in the universe seems to be far less than the amount of mass that is inferred from the interactions of the visible matter. There seems to be far more matter out there than we can see; in fact, the amount of matter implied by the dynamics of the universe on large scales may be even greater than the limits on the amount of hadronic matter in the universe as obtained from the big bang synthesis of nuclei.

Neutrinos with small but nonzero mass might constitute this missing mass, and they could have played a significant role in the formation of galaxies and other structures in the present universe. More speculative particle theories provide other relic candidates for these roles, such as the hypothesized supersymmetric particles and axions. Some limits on the existence of such particles come from astrophysical considerations of the energy flow out of the cores of stars at late stages in their evolution.

In view of these interesting exchanges of information, it is not surprising that cosmologists and particle physicists have been inspired to speculate about much earlier epochs of the big bang, when temperatures and hence particle energies were much higher than present or conceivable particle accelerators could provide. Despite our lack of control of the experimental conditions, the early universe could be a useful laboratory for testing new particle theories. One of the most striking examples has been the realization that grand unified theories that predict baryon decay could also explain the presence of the matter in the universe today. The interactions and decays of superheavy particles with masses above 10^{10} GeV could have assured the present predominance of matter over antimatter. It would no longer need to be assumed as an arbitrary initial condition.

Grand unified theories also predict the existence of magnetic monopoles—potential cosmological relics—which, if they exist, could invalidate current cosmological theories. Their masses of 10^{16} GeV or more are so high that only the early universe could have produced them. Some simple arguments suggest that unacceptably many grand unified monopoles would have been produced in the conventional big bang cosmology, a difficulty that was a stimulus for the proposal of so-called inflationary cosmology. According to this idea, there may have been an early epoch in the history of the universe during which it expanded exponentially, driven by the energy released when there was a change or transition in the state of matter. A large number of particles would be produced when this transition terminated, and the abundance of monopoles would then be greatly diluted. Such an inflationary epoch could also explain many of the greatest cosmological mysteries of the universe, such as the high degree of homogeneity and isotropy that it exhibits and its great age. Inflationary cosmology also enables one for the first time to relate particle physics to the wide range of the large-scale fluctuations in the present universe. It is a challenge to find a particle theory that naturally leads to inflation and to this wide range.

This is one of many areas in which the interaction between particle physics and cosmology will continue to be fruitful in the future.

COSMIC-RAY PHYSICS

Cosmic-ray physics and nuclear physics are the parents of elementary-particle physics. Many of the fundamental discoveries about elementary particles were made using cosmic rays. This is because cosmic rays are a natural source of high-energy particles. Cosmic rays consist primarily of protons that come from outside the solar system. When they hit the atmosphere they make other particles. The positron, the muon, and some of the strange hadrons were discovered and first studied using cosmic rays.

However, with the development of accelerators, the use of cosmic rays in elementary-particle physics has gradually decreased. This is because accelerators provide controllable and much more intense fluxes of particles. It is only the highest energy cosmic rays that are still useful for studies in particle physics—energies that cannot be attained by accelerators.

Thus at present the field of high-energy cosmic rays acts as a bridge between high-energy particle physics and experimental astrophysics. At and above the highest energies reached by hadron-hadron colliders, the energy spectrum, composition, and possible source directions of primary cosmic rays are known to varying degrees. At the same time, the nature of strong interactions at energies above those provided by colliders must be deduced from extrapolations based on known accelerator data and from the largely indirect cosmic-ray data. As the interpretation of these cosmic-ray data in terms of particle-physics phenomenology depends on knowledge of the identity of the initiating cosmic ray (e.g., proton, carbon, or iron nucleus), our knowledge and understanding of both areas are interrelated, and progress is made in an iterative manner as we move to higher energies.

We still do not know the source or acceleration mechanism of high-energy, primary cosmic rays. At the highest observed energies (about 10^{20} eV), it appears that cosmic rays would be too energetic to be trapped in the known magnetic field of our galaxy or to survive energy loss by photoproduction on the relic blackbody radiation in propagation over intergalatic distances. They might come to us from our own local supercluster of galaxies, or they might come from the core of our own galaxy, bent to the Earth by the (unknown) magnetic fields in a galactic halo.

Correspondingly, the only source of information concerning the nature of particle interactions above the highest accelerator energies comes from cosmic rays. Hints of strange, unanticipated phenomena at these energies permeate the cosmic-ray literature. In the past, some cosmic-ray hints, such as evidence for free quarks and monopoles, have not stood up under closer scrutiny. But others, such as the increase of the strong interaction cross section with energy, were later confirmed at particle accelerators.

The problem of studying cosmic rays at energies above 10^{14} eV (greater than those at the CERN proton-antiproton collider) is discussed in Chapter 6. Above 10^{16} eV, the integrated primary cosmic-ray flux is only one per square meter per year.

Other areas addressed by cosmic-ray experiments that overlap astrophysics, particle physics, and nuclear physics include the search for antimatter in cosmic rays and the study of nucleus-nucleus interactions at very high energy. It is quite certain that our local galaxy is composed of ordinary matter, but if antinuclei as heavy as iron are found in primary cosmic rays, even at a level of 10^{-7}, this would be evidence for entire distant galaxies composed of antimatter. Currently there are no data with which to answer this question. Finally, the study of cosmic-ray neutrinos with proton-decay detectors may portend a new field of neutrino astronomy.

NUCLEAR PHYSICS

High-energy physics traditionally is closely linked to an area broadly termed nuclear physics. Elementary-particle physics grew out of nuclear physics; nuclei can be used as probes of elementary particles and vice versa. But the two disciplines are different because they deal with matter at different levels of elementarity.

Accelerators are tools that are common to high-energy and nuclear physics. Low- and medium-energy accelerators used by the nuclear-physics community include meson factories, which produce the most intense beams of protons, pions, and muons, and reactors, which produce the highest fluxes of neutrinos. Some special questions in particle physics must be explored with beams of these kinds.

The nucleus itself is a unique high-density laboratory in which interactions between quarks may be probed, by electron bombardment or by nucleus-nucleus collisions. Conversely, processes such as the production of strange protonlike and neutronlike particles—processes usually associated with particle physics—may be used instead to study nuclear properties, as when the produced strange particle is a nuclear constituent.

An outstanding puzzle is the existence of multiple generations of quarks and leptons. In the case of quarks it is known that the up-down, charm-strange, and bottom-top generations mix, but there is no clear evidence for mixing between members of the electron, muon, and tau lepton families. Searches for muon-electron mixing in muon decay have been carried to exquisite precision with free muons at the LAMPF accelerator at Los Alamos and with muons in the fields of nuclei at TRIUMF (Vancouver) and SIN (Zurich). Mixing of lepton generations in combination with differences in neutrino mass would produce an oscillatory behavior in the composition of neutrino beams. Highly restrictive limits on neutrino oscillations have been set at LAMPF and at reactors in the United States, in France, and in Switzerland, as well as at high-energy physics facilities. Discovery of lepton-generation mixing would be a major achievement, sharpening and expanding our theoretical understanding. Experiments at high sensitivities will continue with this aim.

Successful unification of the strong and electroweak interactions depends on identification of the underlying symmetry group. An outstanding question is the parity symmetry of the electroweak interaction, which at present appears to be fully left-handed. Under the assumption that the neutrino that would participate in any right-handed weak interaction is light enough to be produced in muon decay, the weak-force-carrying particle W_R that would mediate that interaction is required by recent muon-decay data to be at least five times more massive than its left-handed counterpart W_L. If further assumptions including electron nonconservation are made, much more stringent bounds on the W_R mass are set by the limits on neutrinoless double beta decay, e.g., of germanium or selenium isotopes, by direct or geochemical observation. Under the same assumptions, these observations require the electron-neutrino mass to be less than approximately 10 eV. More direct measurement of the electron-neutrino mass is possible through extremely precise study of the endpoint spectrum in tritium beta decay. At present there is unconfirmed evidence from one experiment of finite electron-neutrino mass, in a range that would suggest that neutrinos could account for the dark mass of the universe.

The concept of the quark composition of nucleons has had a major impact on nuclear-physics theory and has led to new ideas in the description of nucleons. For nearly a decade, one direction of research at the Lawrence Berkeley Laboratory Bevalac has been based on the idea that medium-energy heavy-ion central collisions would produce an instantaneous temperature and nuclear density so high that hadronic matter would evolve (deconfine) into an as yet unobserved new state of matter, the quark-gluon plasma. Present estimates indicate that the plasma can be made in two different environments. The first is found at lower energies (a few GeV/nucleon usable energy) in which heavy nuclei are still able to stop in each other, building up high-energy densities in a baryon-rich zone. At much higher energies (above 20-30 GeV/nucleon usable energy), the two colliding nuclei are transparent to each other, leaving a hot baryon-free plasma in the central region after the collision process has taken place. In this central region large densities can be found, sufficient for deconfinement to occur. These large usable energies await the construction of a new accelerator, called by the nuclear science community the Relativistic Nuclear Collider (RNC).

Recently, evidence has indicated that nuclei do not behave simply as a collection of nucleons when high-energy muon, electron, or neutrino scattering occurs. CERN and SLAC experiments find a difference in the form of the quark distribution between deuterium and iron nuclei. This observation will be pursued in the nuclear-physics community through the construction of the SURA 4-GeV electron linear accelerator/stretcher.

These are some examples of the interactions between nuclear physics and elementary-particle physics. Such interactions will continue, not only in the questions that are being studied but also in the accelerators and detectors that are used.

ATOMIC PHYSICS

Effects from new particles are most readily observed at the appropriate energy required to produce the particle. Historically, however, the effects seen in new energy ranges have often been correctly foreshadowed by extrapolation of small deviations from theory observed at lower energies. The extreme precision with which measurements are possible by atomic-physics techniques makes conceivable, even today, the exploration of energy ranges beyond those currently obtained. As an example, atomic experiments were recently used to test the electroweak theory predictions for the synthesis of the weak and electromagnetic theories. These experiments were based on slight differences in the absorption of left- and right-circularly polarized light. Only the extreme precision of laser spectroscopy techniques made these experiments possible.

A second example is the study of two-particle systems that are simpler than the hydrogen atom. The hydrogen atom has at its core an extended, complex object—a proton. In contrast, the positronium system, composed of a positron and electron combined as a short-lived atom, consists of two simple, pointlike particles that exhibit effects that are masked in atomic hydrogen. Muonium, composed of a muon and an electron, is another such simple system that can be formed by stopping muons, produced in an accelerator, in noble gases.

Investigation of the spectra of atomic systems with exotic constituents can also be used to probe particle structure. Examples are provided by atoms composed of muons and pions and of electrons and pions. Deviation from the results expected from pointlike particles provides insights into the structure and interaction of the pion with leptons. Another use of spectroscopy has been the insertion of muons, pions, and kaons into the innermost electron orbits of nuclei to provide, via x-ray spectroscopy, a measure of the electric-field structure in the neighborhood of the nucleus and also to provide high-precision mass measurements of kaons and pions.

A further example is provided by the precise measurements of the magnetic properties of the electron and the muon. The quantum theory of electromagnetism predicts that an electron will act as a small bar magnet and also predicts the strength of that magnet. Atomic-physics experiments have measured that strength accurately and have thus made one of the most careful tests of that theory.

One ingenious small-scale experiment has reported positive evidence for fractional electric charge on small pellets of niobium. An interpretation of these results might be that free quarks on these spheres were responsible for the observations. However, the results have not been confirmed elsewhere, and the consensus in the community is to postpone accepting these results as evidence for free quarks pending strong confirmation.

These classes of experiments investigate properties of matter that are of interest to both atomic and particle physics and are thus an important meeting point for two apparently unrelated areas of physics.

CONDENSED-MATTER THEORETICAL PHYSICS

There has been and continues to be a fruitful and vigorous dialog between theoretical condensed-matter physics and theoretical particle physics. Here we sketch some of the topics in which concepts and techniques of elementary-particle theory enrich condensed-matter physics and also some of the topics where ideas of condensed-matter physics illuminate particle physics.

In the late 1950s, the techniques of quantum field theory used in particle physics started to be employed in condensed-matter physics with outstanding results. Early on, field-theory techniques were used to solve the problem of the energy of an electron gas that forms the starting point for the general discussion of crystals. These techniques were found to be of great importance in dealing with superconductivity and superfluidity. The nature of second-order phase transitions was revolutionized by the use of renormalization group techniques that were developed by both particle and condensed-matter physicists. In turn, the general concepts of phase transitions developed by condensed-matter physicists have been of much use to particle physics in two areas. On one hand, a strongly first-order phase transition has been invoked to produce an inflationary epoch in the early universe that may solve several outstanding cosmological puzzles. On the other hand, there is much recent interest in the possibility of a phase transition in dense, energetic nuclear matter in which the quarks and gluons become deconfined and behave as a quark-gluon plasma. This may happen deep in a neutron star or, perhaps, in high-energy heavy-ion collisions.

It is hoped that the new theory of quantum chromodynamics (QCD) will explain the mass and structure of the hadrons. The method used is to replace the space-time continuum by a discrete lattice of points, and the techniques of calculation employed are quite similar to those first developed in the condensed-matter context. The QCD calculations are of a large scale. The methods are often checked by first applying them to the simpler models of condensed-matter phenomena. Such checks have often proved to shed light on these models. The QCD calculations are so extensive that powerful special-purpose computers are being developed to handle them. This has led to a similar development of special-purpose computers to deal with Ising model calculations in condensed-matter physics.

The concept of order parameters, first introduced in condensed-matter physics, now plays an important role in quantum field theory. We give some examples. These order parameters are akin to the Higgs fields of elementary-particle physics. Liquid helium can flow in a circle about a vortex. Such a flow is analogous to a topological knot, and the vortex line is a kind of defect. Similarly, the Higgs field can arrange itself in a pattern like that of the extended spines of a hedgehog. This is again a type of topological knot. The corresponding defect is a magnetic monopole. Such magnetic monopoles are exceedingly heavy. They may have been produced in the early universe, but they have not yet been detected. Polyacetylene is a long molecule with an alternating bond structure. The bonds can have a jump just as a canal can have an extra lump of water. Such objects are called solitons. A similar situation may occur in the quantum field theory of elementary particles with hadrons being described, at least approximately, as solitons. The total electronic charge about a soliton in polyacetylene is most peculiar—it is half the charge of a free electron. If magnetic monopoles do exist, they would also behave to some extent as solitons, and they could induce a fractional electronic charge.

OTHER APPLICATIONS OF ACCELERATORS

In this section we give some examples of how accelerators have been extended in their applications to other kinds of research and other kinds of technology.

Synchrotron Radiation

The foremost example of the application of accelerators is the use of circular electron accelerators to produce synchrotron radiation. As shown in Figure 7.1, when an electron moves in a circle it emits electromagnetic radiation in a direction tangent to that circle. That electromagnetic radiation covers a broad range of frequencies, extending from the visible to the ultraviolet to the x-ray region of the spectrum. In addition to the broad frequency spectrum, the intensity of the emitted radiation is much higher than can be obtained by other means. For example, the intensity of the x rays within any given frequency range is much greater than can be obtained from a conven-

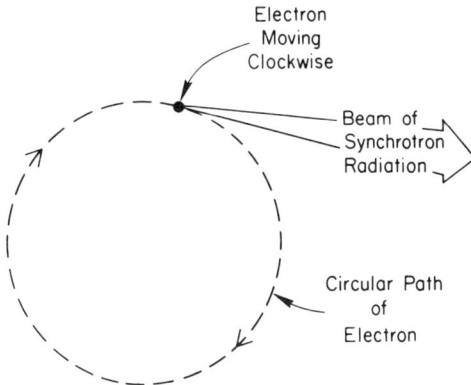

FIGURE 7.1 The most important use of electron storage rings outside of elementary-particle physics is the production of synchrotron radiation. As a high-energy electron moves in a circular orbit it emits an intense beam of x rays called synchrotron radiation. Synchrotron radiation is used for research in many scientific and technical fields: for example, solid-state physics, material science, chemistry, and biology.

tional x-ray tube. This wide frequency spectrum of intense radiation has found many applications in applied physics, material science, electrical engineering, metallurgy, biology, biochemistry, biophysics, and chemistry.

Originally synchrotron radiation was obtained only as a parasitic by-product from circular electron accelerators and from electron-positron colliders. However, as the importance of research based on synchrotron radiation has increased, special dedicated accelerators to produce synchrotron radiation have been built. A list of present-day synchrotron radiation sources now in operation or being constructed is given in Table 7.1. There are more than 20 such facilities.

A simple example of the use of synchrotron radiation has to do with the process called photoionization, in which light is used to eject an electron from an atom or molecule or from a solid. The frequency of the light that ejects the electron tells the researcher something about the structure of the atom, molecule, or solid. Photoionization has been known about since the turn of the century, but to do efficient advanced research at present requires intense sources of light with the frequency involved being known precisely. This is exactly what can be done with synchrotron radiation.

Synchrotron radiation facilities have now begun to develop special kinds of accelerator technology to enhance their capabilities. Synchrotron radiation is produced when an electron goes in a circle, because

TABLE 7.1   Storage Ring Synchrotron Radiation Sources

Location	Ring (lab)	Electron Energy (GeV)	Notes
China			
Beijing	BEPC (IHEP)	2.2–2.8	Parasitic[a]
Hefei	HESYRL (USTC)	0.8	Dedicated
England			
Daresbury	SRS	2.0	Dedicated
France			
Orsay	ACO (LURE)	0.54	Dedicated
	DCI (LURE)	1.8	Partly dedicated
	SuperACO (LURE)	0.8	Dedicated[a]
Germany			
Hamburg	DORIS (DESY)	5.5	Partly dedicated
West Berlin	BESSY	0.8	Dedicated
Italy			
Frascati	ADONE	1.5	Partly dedicated
Japan			
Tsukuba	Photon Factory (KEK)	2.5	Dedicated
	Accumulator (KEK)	6–8	Partly dedicated[a]
Tokyo	TRISTAN (KEK)	30	Parasitic[a]
Okasaki	SOR (ISSP)	0.4	Dedicated
Tsukuba	UVSOR (IMS)	0.6	Dedicated
	TERAS (ETL)	0.6	Dedicated
Sweden			
Lund	Max	0.55	Dedicated
United States			
Gaithersberg, MD	SURF (NBS)	0.28	Dedicated
Ithaca, NY	CESR (CHESS)	5.5–8	Parasitic
Stanford, CA	SPEAR (SSRL)	4.0	Partly dedicated
Stoughton, WI	Tantalus (SRC)	0.24	Dedicated
	Aladdin (SRC)	1.0	Dedicated
Upton, NY	NSLS I (BNL)	0.75	Dedicated
	NSLS II (BNL)	2.5	Dedicated
Soviet Union			
Karkhov	N-100 (KPI)	0.10	Dedicated
Moscow	Kurchatov	0.45	Dedicated
Novosibirsk	VEPP-2M (INP)	0.7	Partly dedicated
	VEPP-3 (INP)	2.2	Partly dedicated
	VEPP-4 (INP)	5–7	Parasitic

[a] Under construction as of April 1985.

any deviation of the path of a particle from a straight line means that the particle is being accelerated. In fact, any means by which an electron can be made to move off a straight line and thus be accelerated will also produce synchrotron radiation. Therefore modern synchrotron radiation accelerators have devices called wigglers or undulators introduced in the path of the electrons. These devices shake the electron as it moves through them, causing strong acceleration and the emission of intense synchrotron radiation in particularly desirable frequency ranges.

Accelerators in Medicine

The electron accelerator is now one of the major tools used in the radiation-therapy treatment of cancerous tissue. The usual way to use such an accelerator is to allow a high-energy beam of electrons to strike a target and then to form the resulting x rays from that target into a narrow, well-defined, and intense beam. The radiation therapist then directs that x-ray beam as carefully as possible onto the tumor that is to be treated. Most treatments involve x rays in the 2- to 6-MeV range, but x-ray energies as high as 30 or 40 MeV are sometimes used. Most of the electron accelerators used for standard radiation therapy are commercially produced linear accelerators. Some circular electron accelerators, based on the betatron principle, are also used.

Although the standard method of radiation therapy using accelerators continues to be the use of x rays, during the last decade there has been a good deal of research on the use of other kinds of particles to destroy cancer. For example, accelerators have been used to produce beams of charged pions, which are then used to treat the tumor. Work has also been done using neutrons and high-energy heavy ions produced in an accelerator.

An interesting new use of accelerators in medicine involves the production of short-lived radioactive materials that produce positrons when they decay. Inside the patient these positrons annihilate, and the resulting photons can then be used in tomography. Because these materials have short lifetimes they cannot be stored but must be produced soon before they are used, and cyclotron accelerators are now being used in hospitals to produce such materials.

High-Intensity Neutron Sources

The scattering of neutrons in mattter has become an important tool in materials science, solid-state physics, polymer chemistry, molecular biology, and other areas of applied and pure science. In the past, nuclear reactors have been the source of the neutron beams used in the scattering experiment. Reactors are still the major source, but spallation neutron sources that use technically advanced proton accelerators are coming into increasing use because they can provide more-intense and higher-energy neutron beams. In a spallation source, a proton beam from a rapid cycling synchrotron bombards a uranium or other heavy-element target, providing a neutron beam.

Accelerators and Plasma Physics

There is an increasing interaction between accelerator physics and technology and plasma physics and technology. This interaction takes several different forms. One example is the use of heavy-ion accelerators to produce inertial fusion. Another example is the use of accelerators to inject charged particles into a plasma to add energy to the plasma as a step toward producing fusion. These ideas are described in detail in the companion volume on plasma and fluid physics.

LARGE-SCALE USES OF SUPERCONDUCTIVITY

The study of superconducting effects and the use of superconducting phenomena play an important part in many areas of physics. Briefly, superconductivity is the absence of electrical resistance that some metals exhibit when cooled to a temperature near absolute zero. That means that an electric current can circulate through the metal without any power loss and therefore could literally circulate forever.

While superconductivity has been used on a laboratory scale for a long time, there have been few large-scale uses of this phenomenon until recently. The most striking example is the recent construction of the Fermilab Tevatron accelerator, which uses about 1000 superconducting magnets. The liquid-helium refrigeration system used to cool those magnets is the largest in the world. In Chapter 5 we discussed the significance of this accomplishment for future construction of very-high-energy proton-proton or proton-antiproton colliders. This accomplishment will also help to lead the way to other large-scale applications of superconductivity.

Large-scale applications of superconductivity require large facilities for cooling and refrigerating with liquid helium, control systems for maintaining the temperature of superconducting devices, and emergency systems for absorbing the sudden power surges that occur if the material suddenly loses its superconducting properties because it warms up. This kind of technology only becomes practical when there has been a sufficient amount of development and engineering work and sufficient experience with big superconducting systems. This is exactly what has been accomplished with the Fermilab superconducting accelerator.

The construction and operation of the 1000-GeV superconducting proton accelerator at Fermilab is the first large-scale use of superconductivity in the world. The technology developed for this accelerator and the experience gained in using it will be useful for other proposed large-scale uses of superconductivity. Some possible applications are listed below:

- Rotating electrical machinery with superconducting windings;
- Superconducting high-power electrical transmission lines;
- Superconducting current-limiting devices for electrical switchgear;
- Superconducting magnet energy storage to smooth peak loads;
- Superconducting coils for separation of materials via their magnetic properties;
- Superconducting magnet systems for fusion reactors;
- Superconducting magnet systems for magnetohydrodynamic power generators;
- Electrodynamic levitation systems for trains using superconductivity.

SUPPORT AND STIMULATION OF NEW TECHNOLOGY

As described in Chapter 6, experiments in elementary-particle physics depend a great deal on the use of integrated circuits, microprocessors, and large high-speed computers. Since the particle physicist uses these devices in an experimental situation, it is often possible to use devices that are not yet fully commercially developed. The researcher will often buy devices or computers that are in the prototype stage in order to have the advantage of using the newest technology. This supports the development of new technology in integrated circuits and in computers.

In addition, there is another valuable effect. The research physicist in elementary-particle physics is often well acquainted with the principles, both physics and engineering, of the new device. Therefore the researcher can often provide information back to the manufacturer about how the prototype device behaves and how it might be improved. Thus elementary-particle physics, through providing for early use of new electronic devices, supports the development of new technology and new devices in electronics and computers.

Superconducting magnet technology is another example. These magnets as used in the Tevatron and, as proposed for use in the Superconducting Super Collider (SSC), use large amounts of superconducting wire. This has stimulated the superconducting metals industry to develop better and cheaper ways for refining and fabrication.

8

Education, Organization, and Decision Making in Elementary-Particle Physics

HISTORICAL BACKGROUND

Before 1960

Before 1940 research in nuclear physics and the construction of accelerators in the United States was carried out at universities and was funded from university general funds, in some cases supplemented by gifts or grants from corporations or individuals. Outside the universities, few industrial and federal research laboratories constructed particle accelerators and carried out research in these areas. Perhaps the most notable research laboratory in the United States was at Berkeley, where E. O. Lawrence had developed the cyclotron and built a sequence of ever larger, more ambitious accelerators.

With World War II and the knowledge of the German discovery of uranium fission, the U.S. nuclear-physics community began several major research and development (R&D) programs funded by the federal government. It is fair to say that big science was born at laboratories such as Los Alamos and Oak Ridge, as well as at large nonnuclear facilities such as the MIT Radiation Laboratory. Projects were accomplished not by one or two senior collaborators assisted by graduate students and skilled technicians, rather a larger group of senior and junior physicists together with professional engineers developed and used large research facilities.

172

After the war, first the Office of Naval Research, then the Atomic Energy Commission, and later the National Science Foundation continued the wartime pattern of federal funding of nuclear science, now again focused at universities. With the discovery of pions in cosmic rays in the late 1940s and the inventions of the betatron, synchrotron, and synchrocyclotron accelerators, a dozen or so major universities built accelerators of over 100-MeV energy to study high-energy nuclear physics. The physicists who implemented these projects applied their experience from the wartime laboratories, and consequently these machines were large, sophisticated engineering undertakings relative to the tabletop experimental equipment of prewar research.

The Berkeley Radiation Laboratory built three large accelerators that became productive research instruments in the late 1940s. A group of East Coast universities meanwhile realized a need to develop a large, cooperative facility, and they joined together to form Associated Universities, Incorporated (AUI). AUI acquired a former army camp on Long Island and developed it into Brookhaven National Laboratory. With funding from the Atomic Energy Commission but operated by AUI, the Laboratory built a 3-GeV proton synchrotron, the Cosmotron, completed in 1953. At Berkeley the 6-GeV Bevatron was completed in 1954, and large liquid hydrogen bubble chambers were developed there, extending the *modus operandi* of big science from the accelerators to the detectors used with them.

During the 1950s, as the complexities of particle interactions and the rich spectra of meson and nucleon states began to unfold, high-energy or elementary-particle physics diverged from nuclear physics and became a distinct field. Although the boundary between these fields remains diffuse, it is appropriate to consider elementary-particle physics as the study of the fundamental constituents of matter and the interactions between them. Nuclear physics, on the other hand, focuses more particularly on the many-body aspects of nuclear forces and nucleon systems.

After 1960 in the United States

During the 1960s, as the press to higher energies required larger accelerators and correspondingly larger detectors and experimental facilities, fewer laboratories became the dominant sites for high-energy research, and the 100- to 400-MeV synchrotrons and cyclotrons on university campuses were phased out. In the 1960s there were about eight accelerators with beam energies greater than 1 GeV in the United States. The largest accelerators at Berkeley, Argonne, Brookhaven,

Cornell, and Stanford were operated by laboratory staff and were used in part by physicists on those staffs. University physicists and their graduate students were major users of these large accelerators and began spending periods ranging from weeks to over a year in residence at the accelerator centers.

Universities evolved research groups of one or more faculty members together with their graduate students, technicians, and postdoctoral research associates to undertake experiments at the national laboratories. Over the past decades these groups have increasingly worked in collaboration with groups from other universities and from the host laboratory.

The funds to support the accelerator laboratories and the university user groups came from the Atomic Energy Commission (AEC), the National Science Foundation (NSF), and the Office of Naval Research (ONR). The support provided by the AEC has continued through its reorganization into the Energy Research and Development Agency (ERDA) and then into the Department of Energy (DOE). The ONR phased out its support in about 1970.

The funding for the university user groups primarily pays for the fabrication of equipment, for travel, and for graduate student stipends. This support has also included salary for faculty members during the summer months, as well as occasionally during the academic year when intensive work on an experiment makes a leave of absence from teaching necessary. This university funding came in the form of research grants (NSF) and contracts (DOE) to the universities, growing in size to over a million dollars per year for some of the large university groups.

The funding for the accelerator laboratories is used for the operation of the accelerators and experimental facilities, for the construction of new equipment and new accelerators, for partial support of the university groups that use the accelerators, and for support for the in-house physics groups that are part of the accelerator laboratory staff. The laboratories also engage in advanced R&D on accelerators and detectors.

In 1965, an advisory group to the AEC recommended the formation of a new national laboratory to build a multi-hundred-GeV proton synchrotron as a national facility and to be operated by a nationally constituted university consortium. Thus in 1966 the Universities Research Association (URA) and the National Accelerator Laboratory [now the Fermi National Accelerator Laboratory (FNAL), or Fermilab] were formed, and an Illinois site was selected for that facility, now the site of the Tevatron.

During the 1970s there were six high-energy accelerator laboratories in the United States serving the elementary-particle physics community: Brookhaven National Laboratory operated by AUI, Fermi National Accelerator Laboratory operated by URA, Lawrence Berkeley Laboratory operated by the University of California, Argonne National Laboratory operated by the University of Chicago, the Laboratory of Nuclear Studies operated by Cornell University, and the Stanford Linear Accelerator Center (SLAC) operated by Stanford University. At present there are four high-energy accelerator laboratories: Brookhaven, Fermilab, Cornell, and SLAC. It may be noted that AUI also operates the National Radio Astronomy Observatory at Greenbank, West Virginia, and the Very Large Array radio telescope at Socorro, New Mexico. The astronomers have emulated the particle physicists and have formed the Associated Universities for Research in Astronomy (AURA), which now operates several astronomical observatories as well as the Space Telescope Science Institute at The Johns Hopkins University.

After 1950 Abroad

The history of accelerator laboratories in Western Europe is similar to that in the United States. In the 1950s and 1960s there were about a half dozen high-energy accelerator laboratories in Europe, located in Great Britain, France, Italy, West Germany, and Switzerland. At present there are two, CERN in Switzerland and DESY in West Germany.

In the middle 1950s European particle physicists joined together to form the European Center for Nuclear Research (CERN) in Geneva, Switzerland. CERN borrowed heavily from the organizational structure of Brookhaven and AUI, and senior American physicists were consulted in developing the organizational structure of this pan-European laboratory and its administration. It was already clear at that time that this field of physics was among the most challenging and exciting of any area of science and that any nation or group of nations wishing to establish scientific leadership must excel in elementary-particle physics. CERN epitomized both that focus of intellectual excitement and a spirit of pan-European cooperation that has proven successful and productive.

In Germany, the Deutsches Elektronen Synchrotron Laboratory (DESY) was established in Hamburg as a focus for particle-physics research. The series of electron accelerators and storage rings constructed there has made major contributions to particle physics over the past two decades, and continues to do so.

The Soviet Union, with some international collaboration, has been active in elementary-particle physics. The Soviets have made major contributions in theory and in research on accelerator physics and technology. Several large accelerators have been built, sometimes at the highest particle energy. They have been less successful in accelerator operation and in exploiting their machines for high-energy physics experiments.

During the last decade, the Japanese, always major contributors to theoretical particle physics, has been developing a major accelerator laboratory called KEK. They have a 12-GeV proton accelerator and are now building an electron-positron collider that will reach about 70 GeV.

At present China is actively entering elementary-particle physics by building an electron-positron collider, called the Beijing Electron Positron Collider (BEPC).

In Western Europe and Japan the organization and funding pattern are similar to those in the United States. The accelerators are located at a few laboratory sites; they are used by physicists from both the universities and the laboratories and the funding is from government sources.

PACE AND PLANNING IN ACCELERATOR CONSTRUCTION AND USE

Most experiments in elementary-particle physics use particle accelerators or colliders; thus these machines lie at the heart of experimental work in this field. The size, complexity, and cost of these machines sets much of the pace and style of research work in this field. The design, construction, and operation of accelerators demands a level of planning and organization that exceeds that required in most other areas of science. It is therefore useful to look at what one might call the life cycle of accelerators.

Conception

The life of an accelerator begins when a group of physicists develops the general conception for a new accelerator. This may be based on a new invention in accelerator technology; for example, the Brookhaven AGS and the CERN PS proton accelerators were based on the invention of alternating-gradient focusing of beams in accelerators. The concept for a new accelerator may also arise because there is a need to go to higher energies or to more intense beams, and there is the realization that existing accelerator technology can be adapted to these new goals. This was the case with the 400-GeV proton accelerator at Fermilab and with the SPS proton accelerator at CERN.

Proposal

The passage from the initial conception of the accelerator to the beginning of its construction requires that a technical design be completely worked out and that the cost of constructing and operating the new accelerator be carefully estimated. This work results in a documented proposal that is submitted to the appropriate government agencies. Thirty years ago this was a relatively simple process; the proposal for the Brookhaven AGS was a six-page letter. In recent years, however, working out the design of a new accelerator has required years of effort and has involved scores of physicists and engineers in the process. The proposal itself is now typically hundreds of pages in length and is backed up by supplemental material in the form of reports from workshops and study groups.

Decision

The proposal is then subjected to a long review process by the government agency involved. Groups inside and outside the agency review the physics justification, the technical soundness, and the cost, and they compare these with competing proposals. For large accelerators this process may include analyses by the legislative as well as executive branch.

Construction

The start of construction of a new accelerator is not always a clear date. Initiation of construction may include acquisition of the land site for the accelerator, the first ordering of materials and supplies, or the setting up of shops and laboratories to begin construction of components. The completion of construction is usually formally marked by the time when the first particle beams are produced. This time is often followed by a period of a year or more during which the accelerator is brought into more efficient operation, the energy of the primary beam is increased, and the intensities of the primary and secondary beams are also increased.

Use of Accelerators for Physics

Outside of the field of elementary-particle physics there is sometimes the notion that an accelerator is built to carry out a certain set of specific experiments and that after those experiments are completed the accelerator is closed down. In fact the situation is very different. Of course the early experiments do carry out the initial goals for which the accelerator was built. But then new physics ideas and new ideas in particle detection lead to experiments on the accelerator for which it may not have been designed. Often the major discoveries made with an accelerator are not those for which it was originally intended. As the accelerator matures, it takes on an even more varied life. Often it is extended once again in energy or in intensity. Sometimes, even more surprisingly, it can be converted into another type of facility. Two examples are the use of the Cornell 10-GeV electron synchrotron as an injector for the CESR electron-positron colliding-beam storage ring and the partial conversion of the CERN SPS proton synchrotron into an extraordinarily successful proton-antiproton colliding-beam storage ring. Other examples are the use of the SLAC linear accelerator as an injector for the SPER and PEP electron-positron storage rings; the use of the DESY 6-GeV electron synchrotron as an injector for the DORIS and PETRA electron-positron storage rings; and the recent conversion of the 400-GeV Fermilab accelerator into an injector for the 1000-GeV superconducting proton ring at Fermilab.

The Death of an Accelerator

Accelerators are shut down when other machines are more effective in carrying out the physics that can be done at that accelerator, or when there are insufficient funds to continue the operation. Appendix A lists most of the major high-energy accelerators built in the United States and in Western Europe during the last 30 years. Perhaps surprisingly, many of these accelerators are still in use. Two examples where lack of funding caused the shut down are the ZGS machine at Argonne and the ISR proton-proton storage ring at CERN. Of the accelerators now in operation, some have had an extraordinary long life. For example, the Bevatron at Lawrence Berkeley Laboratory has been in use for almost 30 years; it is now being used as a heavy-ion accelerator. Another way to measure the usefulness of an accelerator is to see when its major physics discoveries were made. Sometimes, as one would expect, major discoveries occur early in the period of use of an accelerator; for example, the psi (ψ) particle and the tau (τ) lepton were discovered 2 and 3 years, respectively, after the completion of the SPEAR storage ring. Another example is the discovery of the Υ particle at Fermilab 4 years after the 400-GeV accelerator began operation. On the other hand, the *J* particle was discovered at the Brookhaven AGS 14 years after the AGS began operation!

Summary

Thus the life cycle of accelerators spans decades, and the decade is the natural unit to use in thinking about the planning and construction of accelerators. It is also the natural unit for thinking about the pace of experimental research in particle physics and the pace at which new accelerator technologies can be developed. This final point deserves some emphasis. The development of new accelerator technology begins with new ideas such as phase stability, or alternating gradient focusing, or the collision of two beams in a storage ring. But it is a long and difficult path from the new idea to the actual accelerator. Usually the full exploitation of the new idea requires several successive steps in the building of accelerators that go to higher and higher energies or intensities. For example, the concept of a linear accelerator goes back to the late 1920s, but the full use of that idea in the SLAC linac in the 1960s required the building of several smaller linear accelerators in the 1940s and 1950s. Hence, the natural unit of time we use is the decade. This means that our planning must extend over several decades.

THE NATURE OF ELEMENTARY-PARTICLE PHYSICS EXPERIMENTATION

As noted earlier, many experiments in elementary-particle physics are now carried out by large groups of physicists using powerful detectors of large size and complexity. There are exceptions; these include some small-group experiments at high-energy accelerators, at nuclear-physics accelerators, and at reactors and those using cosmic rays. But large-group experiments now dominate, and will continue to dominate, this field. In this section we examine the nature and style of such research.

Large groups including physicists, engineers, and technicians have become necessary because the research apparatus is large and complex. It takes many people to build the experimental apparatus, to maintain it, and to operate it on a 24-hour-a-day basis for months at a time extending over a year or more. If we look more closely at such groups, we see that the cooperative work is made up of a number of coordinated individual activities. The individual nature of the work is especially evident during the design and prototype stages of apparatus construction and also during the data study and analysis stage of the experiment.

During the early stages of the experiment, it is often just one or several physicists who design and build the prototype for a major part of the apparatus, such as a drift chamber or a calorimeter. These physicists are then working in much the same way as physicists working in other fields of research: trying out new ideas in the laboratory, testing new construction techniques, and building prototypes. And this work may include all the traditional skills of the physicist in such areas as mechanical design, electronic design and testing, and fabrication of initial components in the research shop. In such work there is a premium on innovation and improvement of techniques, on simplicity and economy, and on getting the job done right.

The other stage when individual research effort is most important in large-group experiments is at the time of data study and analysis. Almost always just a few physicists, sometimes just one physicist, will concentrate on a particular aspect of physics in the data. For example, in a typical electron-positron collider experiment, different people will be studying different topics such as charm meson or bottom meson physics, or electroweak interference, or searches for new particles. These individuals or small groups tend to carve out a piece of the physics and pursue it on their own. The success or failure of that piece of research depends on the skill and luck of those individuals, just as it does in other areas of science.

The publications that report the results from a large-group experiment are usually signed by the entire group, in recognition of the cooperative effort needed to build and operate the apparatus. But the elementary-particle physics community is relatively small, and within the community it is usually well known who made the leading contribution to the particular piece of physics. Often this is recognized by putting the names of those who did that particular piece of work at the beginning of the list of authors.

A large-group experiment, particularly at a collider, is best looked at as being equivalent to the sum of many different individual experiments of the kind that are carried out at the older fixed-target accelerators. The experimenters have banded together to build one large and complex detector. The price one pays is that there must be a good deal of cooperative work and that it is difficult to rework or rebuild the apparatus quickly. The gain is that the apparatus is very powerful, more powerful than the sum of its parts. Frequently, its power allows

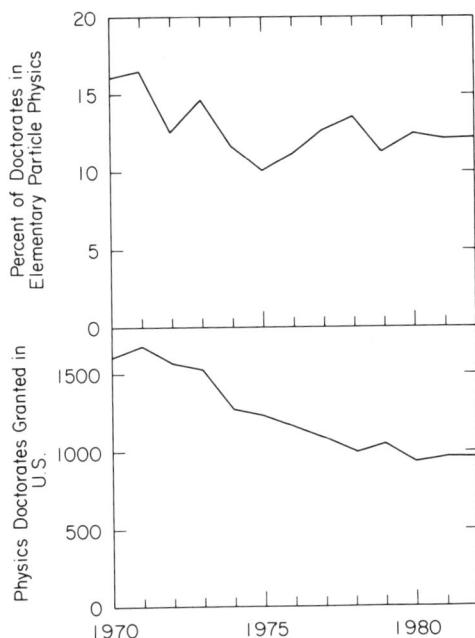

FIGURE 8.1 *Top*: Percentage of physics doctorates granted in the United States that were in elementary-particle physics, either experimental or theoretical. *Bottom*: Number of physics doctorates granted in the United States.

one to do new physics that could not be done by a set of separate and simpler experiments. Indeed, particularly with respect to new particle searches, the large detector permits physics to be done for which one would not have dared to build a special apparatus. Thus it permits speculative physics to be carried out, as well as physics of known phenomena.

GRADUATE EDUCATION

About 1000 doctorates in physics are granted in the United States each year, as shown in Figure 8.1. The physics subfield granting the largest number is solid-state physics; elementary-particle physics ranks second. In 1982, about 12 percent of all physics doctorates granted in the United States were in particle physics, and the dissertations for these degrees were about equally divided between experimental and theoretical physics.

The attractions of elementary-particle physics to the physics graduate student are manifold. Elementary-particle physicists work at the boundary of our knowledge of the nature of matter. Students working on experiments build and use equipment that involves a great range of physical principles and instruments: ionization phenomena in tracking chambers, ultrafast solid-state devices in electronics, high-speed computers, cryogenics systems, and superconducting magnets, for example. Students of theory learn to use new and general theoretical principles such as gauge theories, renormalization group methods, and symmetry breaking. Thus they develop a general problem-solving ability of high order. Elementary-particle physics is an active field, and roughly half of those who are educated in it stay in it. Those who leave the field find excellent uses for their training in other areas.

The graduate education of students in experimental particle physics has been frustrating at times, in view of uncertain experiment scheduling and the occasional breakdown of an accelerator—conditions well beyond the control of the student or his research group. However, the students in research groups gain unusual experience and exposure in other areas. A typical graduate student will complete course work and work on apparatus development at a home university and may then work at a national laboratory for a year or more, setting up, debugging, and collecting data with this equipment. The student then returns to the home university or perhaps continues at the laboratory to carry out the data analysis. Thus the student carries out an individual piece of physics through individual data analysis.

While at the laboratory, students are exposed to an international stream of visitors, seminar speakers, and informal contacts. They have the opportunity to interact with engineers and technicians as well as faculty and students from other institutions and with practicing physicists. The home university and thesis advisor meanwhile continue to provide the continuity and pedagogical foundation around which this broadening experience is molded, exposing the student to the broader range of physics and the other sciences.

INTERACTION BETWEEN THE PARTICLE-PHYSICS COMMUNITY AND THE FEDERAL GOVERNMENT

Universities

The DOE and the NSF support the university users programs. Peer review of research proposals and the alternative of two different agencies have provided a fair and responsive federal structure for the support of university research in this area. New proposals are often submitted to both agencies, and communications between the two sets of Washington physicist-administrators have been good while still maintaining the unique character and perspective of each agency.

An experimental research group must not only attract support from the federal agencies but must also succeed in persuading the program committee advising the accelerator laboratory to allocate accelerator time. Program committees always have members from all parts of the U.S. particle-physics community and often from abroad. This degree of scrutiny of proposals leads to a close filtering of ideas and to a generally high success rate of groups and experiments. The potential liability of this system is that it might tend to choke off unconventional ideas or high-risk explorations. The community is cognizant of this pitfall and has been successful in providing opportunities for exploratory ventures. The Report of the Technical Assessment Committee on University Programs (U.S. Department of Energy, DOE/ER−0182, 1983) to the Division of High Energy Physics, DOE, discusses those points in much more detail.

Accelerator Laboratories

The DOE supports the Brookhaven, Fermilab, and SLAC accelerator laboratories, while the NSF supports the Cornell accelerator laboratory. The work of these laboratories is guided and reviewed in a number of ways by the particle-physics community and by the funding agencies. Each laboratory has a visiting committee that reports to the university body that operates the laboratory. The funding agencies make periodic reviews of the physics research and technology development work of the laboratories. Finally, the High Energy Physics Advisory Panel (HEPAP), discussed below, provides a general overview of the accelerator laboratories. HEPAP's role is particularly important when new accelerator construction is proposed.

At Brookhaven, Fermilab, and SLAC the external university users have formed user organizations. These work with the laboratory administrations on the problems of the visiting physicists and graduate students, as well as on other issues relevant to the research environment and capability of the laboratory.

Decision Making and Advice

Since the end of World War II senior scientists have advised the government in several different ways. The AEC had a General Advisory Committee and, later, under President Eisenhower, a President's

Science Advisory Committee was established. The NSF includes in its advisory structure the National Science Board with members named by the President. In the 1950s a series of decisions related to major new facilities was necessary. An initiative by a group of Midwestern universities to develop a laboratory along the lines of Brookhaven, the desire of the Argonne Laboratory to build a large accelerator, and a Stanford plan for a large electron linear accelerator led the government to seek advice from advisory panels.

In 1967 the AEC formed a standing committee to advise it on the issues it confronts in making decisions in particle physics. This High Energy Physics Advisory Panel (HEPAP) continues to the present. Its 15 members, named for 3-year terms, represent a broad cross section of university and laboratory staff physicists, both theoretical and experimental. The members are named by the Secretary of Energy, with the advice of the DOE director of research for particle physics. HEPAP meets about five times a year. Its agenda is set by the DOE and usually focuses on immediate questions faced by the DOE in particle physics, such as budget issues, program reviews, and international collaboration. HEPAP also appoints subpanels, shown in Table 8.1, to study special questions or broad areas of planning. Its most important decisions relate to the overall direction of the field through its endorsement or rejection of proposals for construction of new facilities. The NSF Program Director for Elementary Particle Physics also regularly attends HEPAP meetings, and the NSF program is included within the purview of this panel. The successful pattern of HEPAP has now been adopted by the nuclear physicists with the formation of the Nuclear Science Advisory Committee (NSAC).

The program committees and user organizations at the major accelerator laboratories have already been mentioned. In addition, the membership of the Division of Particles and Fields (DPF) of the American Physical Society (APS) includes most of the elementary-particle physicists in the United States. Although the DPF has been primarily concerned with planning programs for APS meetings in the past, it now shows promise of becoming more active in policy and planning issues. During the summer of 1982 the DPF organized a 3-week workshop on current questions of particle accelerators, detectors, and physics. The initial planning for the very-high-energy proton-proton collider, the Superconducting Super Collider (SSC), can be traced directly to that meeting. A DPF 3-week workshop in the summer of 1984 was concerned with more detailed planning for the collider.

The European particle-physics community has analogous institu-

TABLE 8.1 Listing of Subpanels of HEPAP

1970	Subpanel on Computer Usage in High Energy Physics
1971	Subpanel on Advanced Accelerator Technology
1970-1972	Subpanel on Future Patterns of High Energy Research
1974	Subpanel on Research and Program Balance
1975	Subpanel on New Facilities
1975	Subpanel on Communicating the Meaning and Accomplishments of High Energy Physics
1975	Subpanel on Requirements of a Vigorous National Program in High Energy Physics
1976	Subpanel on Computing Needs
1977	Subpanel on New Facilities
1978	Subpanel on Study of Impact of Full Cost Recovery on High Energy Physics Community
1978	Subpanel on High Energy Physics Manpower
1979	Subpanel on Accelerator R&D
1980	Subpanel on Review and Planning
1981	Subpanel on Long Range Planning for U.S. High Energy Physics Program
1983	Subpanel on New Facilities
1983	Subpanel on Advanced Accelerator R&D

tions. CERN is governed by a council, consisting of both scientific and political representatives from the CERN member nations. A Scientific Policy Committee advises the CERN Council. In addition, there is a standing European Committee on Future Accelerators (ECFA) that considers long-range planning issues for Europe. The European decision-making process has been generally successful in recent years; the decisions leading to the ISR, the SPS, the proton-antiproton collider, and now LEP have been difficult but are generally agreed to have been timely and correct.

INTERNATIONAL COOPERATION AND COMPETITION

The international nature of elementary-particle physics goes back to the turn of the century. In that period there was no distinction between atomic physics, nuclear physics, and elementary-particle physics, and the great discoveries and advances in those fields came from the nations of Europe. By the 1920s and 1930s, contributions had also begun to come from America and from Asia. The Second World War stopped almost all basic research in Europe and Asia, and in the United States the research establishment was mobilized to develop radar and nuclear and other weapons.

After the war, the United States continued to support substantial research in nuclear physics, as well as in elementary-particle physics as it evolved to become a distinct field. But the destruction caused by the war in continental Europe and in Asia left those regions unable rapidly to resume their traditions in nuclear-physics research. First they had to rebuild their economies and their academic institutions. Thus, for about two decades following the end of the war, substantial progress in particle physics came primarily from the United States and Great Britain. With its greater resources and stronger economy, and aided significantly by its European immigrants, the United States assumed the leadership role in the world in elementary-particle physics research.

By 1960, Europe, Japan, and the Soviet Union had strengthened their economies and had begun to carry out active research in elementary-particle physics. At the same time international cooperation in elementary-particle physics was developing. This cooperation has assumed many forms. The authors and readers of particle-physics journals come from literally dozens of different nations. There have been international meetings and conferences in particle physics every year since 1956. International visits to university and laboratory particle-physics facilities are extensive. Often a physicist will work abroad for several years with a research group in the host country.

There is another form of international cooperation that takes advantage of the moderate to large size of many particle-physics experiments. A group of physicists from one nation can build all or part of an experimental apparatus and take it to another country to use with that country's accelerator. This helps to share the cost of an experiment, makes use of special equipment available in one country, and increases the power of an experiment. American groups have mounted experiments at the CERN and DESY accelerators in Europe. Currently one of the large detectors being built for the LEP electron-positron collider at CERN is directed by an American. Thus far, fewer Western Europeans have come as entire groups to use U.S. accelerators, although Japanese groups have been contributing substantially to experiments at accelerators in the United States and in West Germany. This form of cooperation in the building and operating of detectors is particularly important for the health of the field. Progress in elementary-particle physics depends in the end on successful experiments, and those experiments in turn depend on the quality of the apparatus used. International cooperation helps to improve the quality of the apparatus, while sharing costs. Some international cooperation proceeds informally on a scientist-to-scientist or laboratory-to-laboratory basis, while other efforts are covered by formal intergovernmental agreements. Of course, the outstanding example in our field of an international joint venture is the CERN laboratory in Switzerland. This highly successful laboratory, founded in 1954, is supported by almost all of the nations of Western Europe.

In the future, entire collision regions at colliders might be allocated to foreign groups with some appropriate arrangement for funding and staffing from foreign sources. Given the recent disparity between Western Europe and the United States in the support of new facilities, there is understandably more use by American physicists of European facilities than vice versa.

In all of science, there is some competition along with cooperation. Such competition is necessary for the vigor of science. Competition maintains high standards; it generates diversity of methods and provides cross-checks of experimental findings; and it spurs the scientist to be more inventive, to think harder, and to work harder. Internationally both cooperation and competition exist; the issue is to maintain an appropriate balance between the two.

With respect to elementary-particle physics, the United States had little concern with the right balance between international cooperation and international competition until the last decade. Until the middle 1970s, Western Europe and Japan were still building up their particle-physics research, and the United States led the world of elementary-particle physics. However, this is no longer the case, and we must now consider the balance between cooperation and competition.

The elementary-particle physics community in the United States has developed some guidelines that are intended to maintain this balance:

(a) The continued vigor of elementary-particle physics in the United States requires that there be some forefront accelerator facilities in the United States.

(b) The most productive form of cooperation with respect to accelerator facilities is to develop and build complementary facilities that allow particle physics to be studied from different experimental directions.

(c) The present forms of international cooperation should be continued and supported.

These guidelines are being followed at present. The two accelerator facilities now under construction in the United States are the Tevatron proton-antiproton collider at Fermilab, which will have the highest energy in the world; and the Stanford Linear Collider, which will provide high-energy electron-positron collisions using a new accelerator technology. Western Europe has under construction a higher-energy electron-positron circular collider, LEP, using conventional accelerator technology, and is building an electron-proton collider called HERA. Thus, at the new collider facilities completed or to be completed during this decade (Appendix B), there will be ten experiments (beam-intersection) areas in Europe (two at the CERN $\bar{p}p$ collider, four at LEP, and four at HERA) but only three in the United States (two at the Tevatron and one at SLAC). Therefore, there is now a significant migration of American experimental physicists to exploit the more available European experimental opportunities.

There have been repeated discussions of a truly international accelerator, financed and constructed by a global collaboration. But international cooperation in science, while improving, has not yet reached the point where this appears practical. Questions of the design of the accelerator, of site selection, of funding, and of the allocation of experimental time all appear too unwieldy to be managed by any existing international mechanism. But perhaps most important, the economics of the construction and operation of an international accelerator are not clear. One of the main reasons for international cooperation would be to share the costs, thus reducing the cost borne by each nation. However, the construction and operating cost efficiencies would certainly be decreased in an international effort. For example, the award of construction contracts could not be based solely on lowest bid or best performance, since some consideration would have to be given to spreading the contracts out over the nations contributing to the construction. As another example, design and specifications would become more complicated because of different national technical standards and styles, thus increasing costs and construction time. Thus decreased efficiency would cancel to some extent the hoped-for savings in shared costs. This is a particularly important consideration if the foreign contributions are not large.

There are, however, good reasons for increasing international collaboration beyond the current pattern. Even limited financial contributions of other nations to a new accelerator venture deepens the commitment of all parties. International planning carried out on a nonbinding basis could avoid possible technical mistakes and could help to forge tighter bonds within the international community. The roles of the International Committee on Future Accelerators (ICFA) and of the Summit Working Group on High Energy Physics have recently been strengthened in this respect. We welcome these valuable additions.

Thus the time is not yet ripe for a truly global collaboration. Through the next generation of accelerators, including the proposed very-high-energy proton-proton collider, the SSC, it seems sensible to retain the primary funding, the governance, and the management of the SSC in the United States. International help and cooperation should be sought in providing some of the experimental facilities and possibly some of the construction cost. The management should ensure that the accelerator is open to the entire international particle-physics community and that mechanisms for collaboration with non-U.S. physicists and research teams are developed and encouraged. But the U.S. elementary-particle physics community, working with the federal government, must assume the primary responsibility for initiating and building this accelerator.

FUTURE TRENDS AND ISSUES

In the final section of this chapter we describe some of the future trends that we perceive in the organization and education associated with elementary-particle physics. We also discuss some issues that may arise and make some recommendations aimed at resolving those issues.

Graduate Students' Role

Particle physics has always been characterized by an infectious intellectual excitement, and this is currently being fueled by remarkable advances in our understanding of elementary particles. While this continues to attract good students into the field, the appeal of a Ph.D. thesis research program in experimental particle physics is tempered by the potential for a long and uncertain schedule and by the perception of an impersonal relationship as a member of a large team. As with every field of science, the future vitality of the field is critically dependent on the quality of young people who enter as graduate students and constitute the young Ph.D.s. The particle-physics community must strive to maintain modalities that will make it possible for graduate students to play a significant, creative role in these large experiments and to complete a Ph.D. thesis in a reasonable time. Basically, as discussed above, graduate students must continue to have the opportunity to carve out specific pieces of physics for their own research.

Scientific Manpower in Particle Physics

The demographics of the field should be well understood. The quality and quantity of the graduate-student influx into particle physics, the dispersion of particle physicists into other areas, and the division between theory and experiment should be known and monitored. The particle-physics community has remained at nearly a constant size in spite of producing new Ph.D.s at a rate of several percent of its total per year. About half or more of the particle-physics Ph.D.s use their education to move into fields as diverse as astronomy, fusion research, computer science, and nuclear medicine. Because particle physics is entirely basic research with no direct applied aspects, no industrial laboratories maintain significant particle-physics programs, and the field exists entirely within the universities and the national laboratories.

As university undergraduate enrollments shrink between now and the end of the century the universities will be able to justify fewer faculty positions, and as particle physics is a young field, relatively few faculty in this area will retire soon, as is apparent in Figure 8.2. It thus may be necessary to fund through federal grants and contracts increasing numbers of research faculty and research scientist appointments in particle physics at universities in order to maintain the youth and vitality of the university programs. There is some evidence of a trend in this direction; it should be understood, monitored, and supported. Competent young scientists should be able to perceive a clear career ladder in the universities as well as in the national laboratories.

Advanced Accelerator and Detector Research

It is clear from Chapter 5 that the particle-physics community has invested substantial effort and ingenuity in the invention and development of particle accelerator systems over the past 50 years. Correspondingly, the future progress of the field depends on the continuation of this trend. With the concentration of elementary-particle physics accelerators into only four laboratories in the United States, and with only two of these at universities, few graduate students are being educated in the physics of particle accelerators. In the programs of the large laboratories there is generally some provision for work in advanced accelerator research. But often, when budget reductions occur, this research may be sacrificed in favor of maintaining a strong experimental research program and the momentum of construction of authorized new facilities.

A method should be developed to educate young physicists in accelerator theory and to support in a consistent manner long-range research in particle accelerators. This is essential not only for the long-range future of particle physics: accelerator physics is a significant

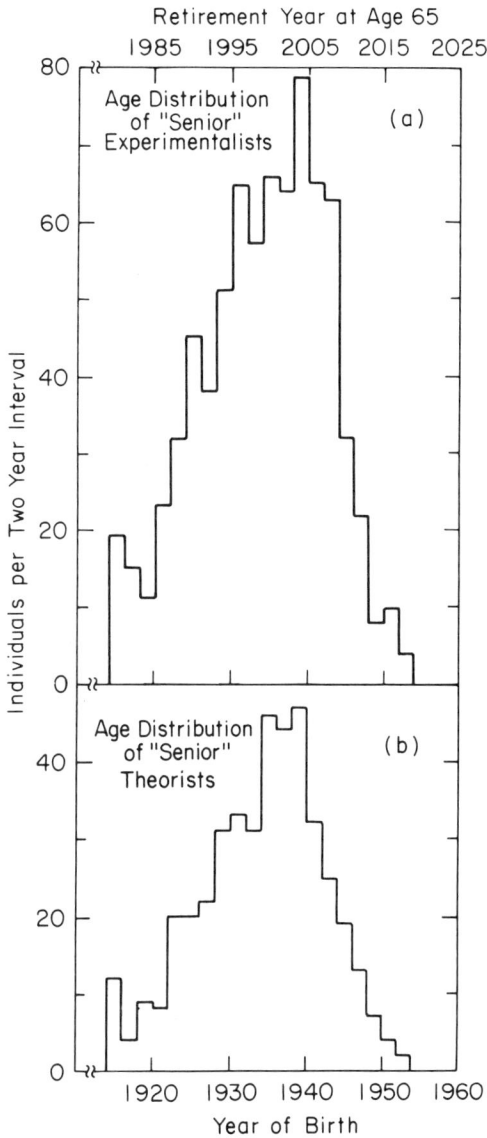

FIGURE 8.2   Age distribution of senior experimentalists and theorists in elementary-particle physics. Senior means associate and full professors and laboratory equivalent. (Report of the Technical Assessment Committee on University Programs, 1983.)

192 ELEMENTARY-PARTICLE PHYSICS

area where particle physics overlaps other fields, and the spinoff from accelerator physics to other fields has been particularly valuable. There is no reason that future accelerator research and development should not continue this trend.

Similar remarks are appropriate with reference to detector developments. Although advances in detector concepts and design are still dispersed among the universities as well as the laboratories, there is a trend here as well to reduce this effort and to concentrate it at a few national laboratories. It remains true that an advance in detector technique can be equivalent to an improvement in accelerator beam intensity or luminosity in the study of new phenomena. Encouragement and support of detector development, at the universities as well as at the large laboratories, should continue.

Laboratory Management

The particle-physics community has been comfortable with the management of the large laboratories by universities, either singly or in consortia. There is no motivation to change this arrangement. If a new laboratory is created around the SSC, it might best be managed similarly, most probably by a national consortium of universities. The management by universities or university consortia of the large accelerators of today—facilities costing in excess of a hundred million dollars—has resulted in an enviable record in terms of meeting goals of performance, budget, and schedule. One might question whether the scale of the SSC is so far beyond our current experience that an industrial management group, familiar with the implementation of very large high-technology projects for the government, might be a better alternative. Yet there is no evidence that performance by industry in major space projects, reactor construction projects, or large highly technical military systems has been superior; if anything there is evidence in the opposite direction. Moreover, the particle-physics community is in favor of university management, and a strong case would need to be made for an alternative. The basic research in particle physics, even on the Olympian scale of the SSC, will have scholarly academic goals, and the SSC must be managed to maintain this focus. University management furthermore buffers the laboratory from political and commercial motivations that might enter under other management structures.

One change in past practice that could be considered for a new laboratory would be the limitation of a director's tenure to 5 (or so) years, as is the case at CERN. Although such a policy for the existing laboratories might also be desirable, the responsibility for such a change must rest with the managements of the respective laboratories. A 5-year term would have the advantages of maintaining leadership vitality and of encouraging productive scientists to accept a directorship without the implication of a commitment for the duration of a professional career. Alternatively, a 3- or 4-year term, renewable once only, might be considered.

Advisory Structure

HEPAP has been generally successful. This kind of peer input into the federal decision-making process is obviously effective.

The frequent convening of ad hoc panels to consider long-range planning issues and other specific questions is evidence that community input beyond that of HEPAP is also important. There has been occasional discussion about establishing a standing long-range planning committee in the United States, analogous to ECFA in Europe, but there is no consensus on this question. It appears that the Division of Particles and Fields of the American Physical Society will become increasingly active through its organization of workshops and studies, and these will contribute significant community input to the decision-making process. It is in any event most desirable to continue to examine and improve the planning mechanisms for high-energy physics.

9

Conclusions and Recommendations

THE REVOLUTION OF THE PAST TWO DECADES

During the past two decades our understanding of the fundamental nature of matter has undergone a revolution. We have found three families of elementary particles: the family of leptons, the family of quarks, and the family of force-carrying particles. The lepton family and the quark family each consist of three generations, with strong similarities between the generations.

The four fundamental forces—electromagnetic, gravitational, strong, and weak—were known earlier, but during the last two decades the electromagnetic and weak forces have been unified theoretically and experimentally. In addition, the particles carrying the strong force, called gluons, and the particles carrying the weak force, the W and Z, have each been discovered.

During the same period, the vast family of particles called the hadrons has been shown not to be elementary in themselves, but rather to be made up of quarks. In particular, the quark nature of the proton and neutron, the most common hadrons, has been studied and understood in great detail. We have also acquired a good understanding of how quarks behave inside hadrons and of how quarks interact when hadrons are involved in high-energy collisions.

194

HOW THE REVOLUTION WAS MADE

This revolution in elementary-particle physics came about because of three interacting components of particle research:

Progress in Accelerators. Particle colliders and higher energy fixed-target accelerators have come into operation during the last two decades. These new facilities provided the high-energy particles that were needed for most of the experimental work involved in this revolution.

New Experiments and New Experimental Techniques. In physics, new discoveries are made experimentally, and all new ideas and theories must be tested and validated experimentally. Some of the revolution in particle physics occurred because theory predicted the existence of new phenomena or new particles. That is how the *W* was discovered. Conversely, sometimes experiments led the way. The tau lepton was discovered through an experimental search, and thus the concept of the third generation was introduced. Almost all of these experiments used new experimental techniques such as particle-detecting wire chambers and integrated circuitry.

Theoretical Progress. The third necessary component of elementary-particle research is progress in elementary-particle theory. Sometimes that progress is in the form of an elegant mathematical theory, such as the theory that unifies the weak and electromagnetic forces. Sometimes it is in the form of a broad insight; for example, the realization that most of the properties of hadrons can be directly explained by models of the behavior of quarks inside the hadrons.

WHAT WE WANT TO KNOW

With this revolution accomplished, we are now led to deeper questions about the basic nature of matter and energy. These questions could not be asked in a sensible way until we had identified the three families of elementary particles and learned how the four basic forces behave. Some of these questions express our dissatisfaction with present theories, which require several dozen unexplained numerical constants. What sets the values of these constants? Are they interconnected or independent? Among these constants are the masses of the various elementary particles and the strengths of the various basic forces. The masses of the particles are completely unexplained, because we do not yet understand the origin of mass.

Other questions that we need to answer include the following: How many generations of quarks or leptons are there? Are the quarks and leptons related to each other? If so, how? Can the strong force be unified with the already unified electromagnetic and weak forces?

Then there are the questions that are related to our overview of elementary-particle physics. Are the quarks and leptons really elementary? Are there yet other types of forces and elementary particles? Can gravitation be treated quantum mechanically as are the other forces, and can it be unified with them? More generally, does quantum mechanics apply in all parts of elementary-particle physics? Do we understand the basic nature of space and time?

These questions indicate the hopes and opportunities for continued progress in elementary-particle physics during the next several decades. Although the United States has until quite recently been the leading contributor to elementary-particle physics research, gradually other regions, particularly Western Europe and Japan, have substantially increased their contributions to this research. This is as it should be, since science is a worldwide endeavor. Elementary-particle physics is a basic science. It interacts with many other areas of physics and astronomy, and it develops, stimulates, and uses many new technologies. In the belief that the United States should maintain a forefront role in particle-physics research, we conclude this report with a set of recommendations for the elementary-particle physics program in the United States. No priority is indicated by the order in which we present these recommendations. This program has as its centerpiece the plan, based on ongoing accelerator development and design work, for the construction in the United States of a very-high-energy, superconducting proton-proton collider, the Superconducting Super Collider (SSC).

In arriving at this program the elementary-particle physics community has had to choose from a number of alternative physics directions and from proposals for building other types of new accelerators and facilities. A great deal of thought, research, and discussion has gone into the program described here.

RECOMMENDATIONS FOR UNIVERSITY-BASED RESEARCH GROUPS AND USE OF EXISTING FACILITIES IN THE UNITED STATES

The community of elementary-particle physicists in the United States consists of about 2400 scientists, including graduate students, based in nearly 100 universities and 6 national laboratories. They work together in groups frequently involving several institutions. It is their experiments, their calculations, their theories, their creativity that are at the heart of this field. The diversity in size, in scientific interests, and in styles of experimentation of these research groups are essential for maintaining the creativity in the field. *Therefore we recommend that the strength and diversity of these groups be preserved.*

Most elementary-particle physics experiments in the United States are carried out at four accelerator laboratories. Two fixed-target proton accelerators are now operating: the 30-GeV Alternating Gradient Synchrotron (AGS) at the Brookhaven National Laboratory and the 1000-GeV superconducting accelerator, the Tevatron, at the Fermi National Accelerator Laboratory. Cornell University operates the electron-positron collider CESR. The Stanford Linear Accelerator Center operates a 33-GeV fixed-target electron accelerator, which also serves as the injector for two electron-positron colliders, SPEAR and PEP. In addition, some elementary-particle physics experiments are carried out at medium-energy accelerators that are primarily devoted to nuclear physics.

Experimentation at the four accelerator laboratories requires complex detectors that are often major facilities in their own right. The equipment funds for major detectors and the operating funds for the accelerators have been insufficient to allow optimum use. Because accelerator laboratories necessarily have large fixed costs, the productivity of the existing accelerator facilities can be increased considerably by a modest increase in equipment and operating funds. *We recommend fuller support of existing facilities.*

RECOMMENDATIONS FOR NEW ACCELERATOR FACILITIES IN THE UNITED STATES

The capability of two existing accelerators in the United States is now being extended by adding collider facilities to each of them. A 100-GeV electron-positron collider, using a new linear collider principle, is now being constructed at the Stanford Linear Accelerator Center. The Tevatron at the Fermi National Accelerator Laboratory is being completed so that the superconducting ring can also be operated as a 2-TeV proton-antiproton collider. *We recommend continued support for the completion of these new colliders on their present schedule. In addition, we recommend that their experimental facilities and programs be fully developed.*

The United States elementary-particle physics community is now carrying out an intensive research, development, and design program intended to lead to a proposal for a very-high-energy, superconducting proton-proton collider, the Superconducting Super Collider (SSC). It will be based on the accelerator principles and technology that have been developed at several national laboratories, in particular the extensive experience with superconducting magnet systems that has been gained at the Fermi National Accelerator Laboratory and Brookhaven National Laboratory. The SSC energy would be about 20 times greater than that of the Tevatron collider. This higher energy is needed in the search for heavier particles, to find clues to the question of what generates mass, and to test new theoretical ideas. Our current ideas predict a rich world of new phenomena in the energy region that can be explored for the first time by this accelerator. Furthermore, history has shown that the unexpected discoveries made in a new energy regime often prove to be the most exciting and fundamentally important for the future of the field. On its completion this machine will give the United States a leading role in elementary-particle physics research. *Since the SSC is central to the future of elementary-particle physics research in the United States, we strongly recommend its expeditious construction.*

RECOMMENDATIONS FOR ACCELERATOR RESEARCH AND DEVELOPMENT

Since accelerators are the heart of most elementary-particle experimentation, physicists are continuing research and development work on new types of accelerators. Indeed, technological innovation in accelerators has been the driving force in extending the reach of high-energy physics. An important part of this work is concerned with extending the electron-positron linear collider to yet higher energies. One of the purposes of the construction of the Stanford Linear Collider is to serve as a demonstration and first use of such a technology. Advanced accelerator research is also exploring new concepts, based on a variety of technologies, that may provide the basis for even more powerful accelerators, perhaps to be built in the next century. Such research also leads to advances in technology for accelerators used in industry, medicine, and other areas of science such as studies based on synchrotron radiation. *We recommend strong support for research and development work in accelerator physics and technology.*

RECOMMENDATIONS FOR THEORETICAL RESEARCH IN PARTICLE PHYSICS

Theoretical work in elementary-particle physics has provided the intellectual foundations that motivate and interconnect much experimental research. Theorists working in elementary-particle physics have also played an important role in forging links with other disciplines, including statistical mechanics, condensed-matter physics, and cosmology. Theoretical physicists make vital contributions to university research programs and to the education of students who will enter all branches of physics.

We recommend that the existing strong support for a broad program of theoretical research in the universities, institutes, and national laboratories be continued. A new element of theoretical research is the increasing utilization of computer resources, which has spurred the development and implementation of new computer architectures. This trend will require the evolution of new equipment-funding patterns for theory.

RECOMMENDATIONS FOR NONACCELERATOR PHYSICS EXPERIMENTS

It is appropriate that some fraction of the particle-physics national program be devoted to experiments and facilities that do not use accelerators. These experiments include the searches for proton decay using large underground detectors, the use of cosmic rays to explore very-high-energy particle interactions, the measurements of the rate of neutrino production by the Sun, and the use of nuclear reactors to study subtle properties of neutrons and neutrinos. There are also diverse experiments searching for evidence of free quarks, magnetic monopoles, and finite neutrino mass. Still other classes of experiments overlap the domain of atomic physics; these include exquisitely precise tests of the quantum theory of electromagnetism, studies of the mixing of the weak and electromagnetic forces in atomic systems, and searches for small violations of fundamental symmetry principles through a variety of different techniques. Many of these are small-scale laboratory experiments. Some provide a means of probing an energy scale inaccessible to present-day accelerators.

The value of these experiments is substantial. *They will continue to play a vital role that is complementary to accelerator-based research, and we recommend their continued support.*

RECOMMENDATIONS FOR INTERNATIONAL COOPERATION IN ELEMENTARY-PARTICLE PHYSICS

Our program should be designed to preserve the vigor and creativity of elementary-particle physics in the United States and to maintain and extend international cooperation in the discipline. *We recommend four guidelines for such a balanced program.* First, the continued vitality of American elementary-particle physics requires that there be forefront accelerator facilities in the United States. The use of accelerators developed by other nations provides a needed diversity of experimental opportunities, but it does not stimulate our nation's technological base as does the conception, construction, and utilization of innovative facilities at home. The Superconducting Super Collider will be a frontier scientific facility, and the technological advances stimulated and pioneered by its design and construction will serve the more general societal goals as well. Second, the most productive form of cooperation with respect to accelerators is to develop and build complementary facilities that allow particle physics to be studied from different experimental directions. Third, the established forms of international cooperation, including the use of accelerators of one nation by physicists from another nation, should be continued. Fourth, looking beyond the program proposed in this report, there should be further expansion of international collaboration in the planning and building of accelerator facilities.

CONCLUSION

We believe that the implementation of these recommendations will enable the United States to maintain a competitive forefront position in elementary-particle physics research into the next century. Central to this future is the construction of the SSC, the very-high-energy proton-proton collider using superconducting magnets.

Appendixes

A

The World's High-Energy Accelerators

This Appendix lists high-energy accelerators that are now in operation or that have operated within the past few years. High energy is defined for fixed-target accelerators as a primary-beam energy greater than 5 GeV and for colliders as a total energy greater than 3 GeV.

TABLE A.1 Fixed-Target Proton Accelerators

Name	Type	Maximum Energy (GeV)	Location	Year Construction Completed	Present Use or Year Closed	Years Used
Bevatron	Weak focusing	6	LBL, U.S.A.	1954	Converted to heavy ion accelerator; still in use	>30
Nimrod	Weak focusing	7	Rutherford, U.K.	1963	Closed in 1978	15
ZGS	Weak focusing	12	Argonne, U.S.A.	1963	Closed in 1979	16
KEK	Strong focusing	12	Tsukuba, Japan	1976	In use for elementary-particle physics	>8
PS	Strong focusing	28	CERN, Switzerland	1959	In use for elementary-particle physics and as injector	>25
AGS	Strong focusing	33	Brookhaven, U.S.A.	1960	In use for elementary-particle physics	>24
76-GeV Proton Synchrotron	Strong focusing	76	Serpukhov, USSR	1967	In use for elementary-particle physics	>17
SPS	Strong focusing	400	CERN, Switzerland	1976	In use for elementary-particle physics and as collider	>8
400-GeV Proton Synchrotron	Strong focusing	400	Fermilab, U.S.A.	1972	In use as injector for Tevatron	>12
Tevatron	Strong focusing, superconducting	1000	Fermilab, U.S.A.	1983	In use for elementary-particle physics; will be used as collider	>1

TABLE A.2 Fixed-Target Electron Accelerators

Name	Type	Maximum Energy (GeV)	Location	Year Construction Completed	Present Use or Year Closed	Years Used
6-GeV Electron Synchrotron	Circular	6	Harvard/ MIT, U.S.A.	1962	Closed 1973	11
6-GeV Electron Synchrotron	Circular	7	DESY, Germany	1964	In use as injector for storage rings	>20
12-GeV Electron Synchrotron	Circular	12	Cornell, U.S.A.	1967	In use as injector for CESR storage ring	>17
2-Mile Linear Accelerator	Linear	33 being raised to 50	SLAC, U.S.A.	1966	In use for elementary-particle physics and as injector for storage rings	>18

TABLE A.3 Hadron-Hadron Storage Ring Colliders

Name	Type	Maximum Energy (GeV)	Location	Year Construction Completed	Present Use or Year Closed	Years Used
ISR	Proton –proton	62	CERN, Switzerland	1971	Closed in 1983	12
$S\bar{p}pS$ collider	Proton –antiproton	~600	CERN, Switzerland	1982	In use for elementary-particle physics	>2

TABLE A.4   Electron-Positron Storage Ring Colliders

Name	Type	Maximum Energy (GeV)	Location	Year Construction Completed	Present Use or Year Closed	Years Used
SPEAR	Two interaction regions	8	SLAC, U.S.A.	1972	In use for elementary-particle physics and as synchrotron light source	>12
DORIS	Two interaction regions	10	DESY, Germany	1973	In use for elementary-particle physics, as synchrotron light source, and as injector	>11
VEPP-4	Two interaction regions	14	Novosibirsk, USSR	1979	In use for elementary-particle physics	>5
CESR	Two interaction regions	16	Cornell, U.S.A.	1979	In use for elementary-particle physics and as synchrotron light source	>5
PEP	Six interaction regions	36	SLAC, U.S.A.	1980	In use for elementary-particle physics	>4
PETRA	Four interaction regions	46	DESY, Germany	1978	In use for elementary-particle physics	>6

B

Particle Colliders Under Construction

Table B.1 lists the particle colliders now under construction in the world. Of particular interest for very-high-energy experiments is the number of interaction regions available for experiments at colliders with sufficient energy to study the new physics of the W and Z particles. There will be eight such regions in Western Europe, not including the two already in use at the CERN proton-antiproton collider. There will be three such regions available in the United States.

207

208

TABLE B.1 Particle Colliders Under Construction

Type	Name	Place	Total Energy or Beam Energies	Number of Interaction Regions	Sufficient Energy for Studying New Physics W and Z
Electron-positron	BEPC	China	5.6 GeV	2	No
Electron-positron	TRISTAN	Japan	70 GeV	4	No
Electron-positron	SLC	United States	100 GeV (initial) 140 GeV (maximum)	1	Yes
Electron-positron	LEP	Western Europe	100 GeV (initial) 200 GeV (maximum)	4	Yes
Electron-proton	HERA	Western Europe	30-GeV electrons + 820-GeV protons	4	Yes
Proton-proton	UNK	Soviet Union	400-GeV protons + 3000-GeV protons	?	Yes
Proton-antiproton	Tevatron I	United States	2000 GeV	2	Yes

C

Statistical Information on Elementary-Particle Physics Research in the United States

PHYSICISTS AND GRADUATE STUDENTS IN ELEMENTARY-PARTICLE PHYSICS

Table C.1 gives the number of Ph.D.-level physicists doing elementary-particle physics research in the United States. The total increases to about 2400 when the number of graduate students doing doctoral research in elementary particles is added in. Figure C.1 shows where the Ph.D.-level physicists work.

Table C.2 gives the number of doctorates in elementary-particle physics granted per year in the United States. The number of elementary-particle physics doctorates is compared in Figure C.2 with all physics doctorates and with all physical science doctorates. These personnel data come from the Report of the Technical Assessment Committee on University Programs, DOE Report DOE/ER−0182 (1983).

FUNDING FOR ELEMENTARY-PARTICLE PHYSICS RESEARCH

Elementary-particle physics research is supported by the Department of Energy and by the National Science Foundation in the United States. Figure C.3 shows the total funding for elementary-particle physics in the United States since 1967. The funding is corrected for inflation and expressed in equivalent 1984 dollars.

TABLE C.1   Ph.D.-Level Research Personnel in U.S. Elementary-Particle Physics[a]

	Universities		Laboratories		Subtotal		
Date	Theor.	Exptl.	Theor.	Exptl.	Theor.	Exptl.	Total
5/27/68	498	563	29	399	527	962	1489
10/22/70	576	693	60	391	636	1084	1720
1/01/73	501	704	84	391	585	1095	1680
1/01/75	528	741	76	379	604	1120	1724
1/01/78	562	731	89	448	651	1179	1830
1/01/81	631	798	108	534	739	1332	2071

[a] AEC/ERD/DOE Census Data.

TABLE C.2   Doctorates Granted in the United States in Elementary-Particle Physics

Academic Year	Ph.D.s in Particle Physics[a]	Fraction of Experimentalists[b]
1969-1970	256	0.47 ± 0.04
1970-1971	277	0.47 ± 0.05
1971-1972	198	0.47 ± 0.04
1972-1973	222	0.55 ± 0.05
1973-1974	146	0.43 ± 0.05
1974-1975	126	0.47 ± 0.06
1975-1976	130	0.52 ± 0.06
1976-1977	138	0.50 ± 0.06
1977-1978	135	0.58 ± 0.07
1978-1979	121	0.59 ± 0.06
1979-1980	117	0.59 ± 0.06
1980-1981	117	0.48 ± 0.05

[a] NRC, Doctoral Records File.
[b] AIP, Graduate Student Survey.

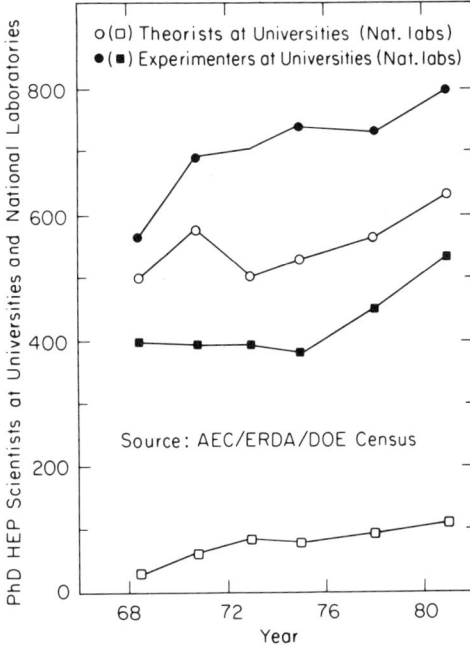

FIGURE C.1 Locations of Ph.D.-level elementary-particle physicists in the United States.

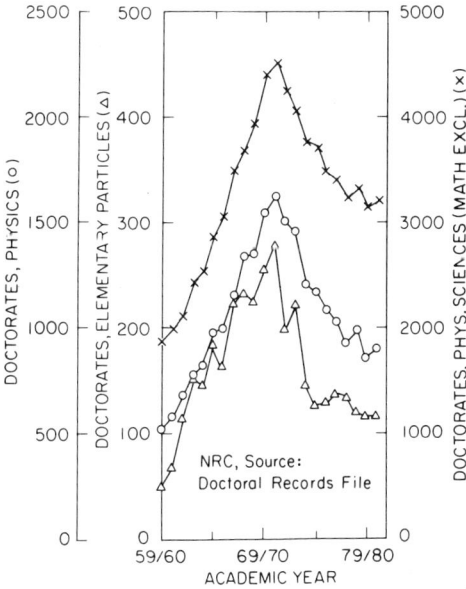

FIGURE C.2 Number of doctorates per year granted in elementary-particle physics, in all of physics, and in the physical sciences in the United States.

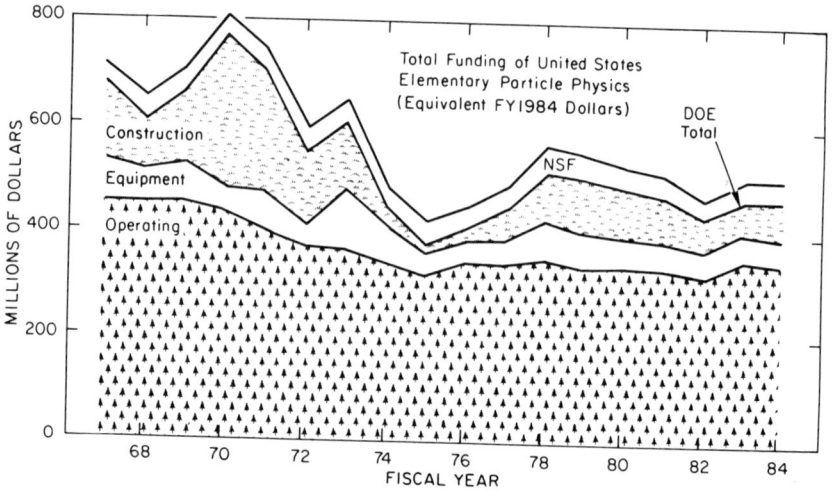

FIGURE C.3   Funding per year for elementary-particle physics in the United States in terms of equivalent FY 1984 dollars.

Glossary

Accelerator. A device that increases the energy of charged particles such as electrons and protons.

AGS. The Alternating Gradient Synchrotron, a 33-GeV proton accelerator at the Brookhaven National Laboratory.

Annihilation. See *Antiparticle.*

Antimatter. Matter composed of antiparticles, i.e., antiprotons, antineutrons, antielectrons, instead of, i.e., the ordinary protons, neutrons, electrons.

Antiparticle. Each particle has a partner, called an antiparticle, which is identical except that all chargelike properties (electric charge, strangeness, charm, for example) are opposite to those of the particle. When a particle and its antiparticle meet, these properties cancel out in an explosive process called annihilation. The particle and antiparticle can then disappear and other particles be produced.

Antiproton. The antiparticle partner of the proton.

Astrophysics. Physics applied to astronomy and astronomical phenomena such as the evolution of stars and the formation of galaxies.

Asymptotic freedom. The concept that the strong force between quarks gets weaker as the quarks get close together.

Atom. The smallest unit of a chemical element, approximately 1/100,000,000 centimeter in size, consisting of a nucleus surrounded by electrons.

Baryon. A type of hadron. The baryon family includes the proton,

213

neutron, and those other particles whose eventual decay products include the proton. Baryons are composed of three-quark combinations.

Beam.   A stream of particles produced by an accelerator.

Beauty.   See *Bottom.*

BEPC.   A circular electron-positron collider with a total energy up to 6 GeV and high luminosity, under construction near Beijing, China.

Beta decay.   The decay of a particle or nucleus by the emission of an electron or positron through the weak interaction.

Bevatron.   A circular accelerator at the Lawrence Berkeley Laboratory, Berkeley, California; previously used to accelerate protons up to 6 GeV and now part of a complex for accelerating nuclei.

BNL.   Brookhaven National Laboratory.

Bottom.   The distinguishing characteristic of the fifth type of quark, also called the *b* quark or beauty quark. Each quark is characterized by a number of properties, including familiar ones like mass and electric charge, and less familiar ones that were arbitrarily given names like bottom and charm.

Broken symmetry.   The failure of a symmetry principle owing to the presence of an additional force or phenomenon.

Bubble chamber.   A particle detector in which the paths of charged particles are revealed by a trail of bubbles produced by the particles as they traverse a superheated liquid. Hydrogen, deuterium, helium, neon, propane, and Freon liquids have been used for this purpose.

Calorimeter.   A particle detector in which the total energy carried by a particle or group of particles is measured.

Cerenkov counter.   A detector of Cerenkov radiation, which is electromagnetic radiation emitted by a charged particle when it passes through matter at a velocity exceeding that of light in that material.

CERN.   The European Center for Nuclear Research, located near Geneva, Switzerland, and supported by most of the nations of Western Europe.

CESR.   The Cornell Electron Storage Ring, an electron-positron collider with a maximum total energy of 16 GeV located at Cornell University.

Charm.   The distinguishing characteristic of the fourth type of quark, also called the *c* quark. Each quark is characterized by a number of properties, including familiar ones like mass and electric charge, and less familiar ones that were arbitrarily given names like charm and bottom.

Charmonium.   The family of hadronic particles composed of a charm quark and an anticharm quark.

Circular accelerator. An accelerator in which the particles move around a circle many times, being accelerated further in each revolution around the circle.

Collider. When a high-energy particle collides with a stationary target, a large portion of the energy resides in the continuing forward motion. Only a small portion of the energy is available for creating new particles. In a collider, collisions take place between high-energy particles that are moving toward each other. In such an arrangement, most of the energy is available for creating new particles.

Colliding-beam accelerator. See *Collider.*

Color. A property of quarks and gluons, analogous to electric charge, which describes how the strong force acts on a quark or gluon.

Conservation law. A physical law that states that some quantity or property cannot be changed in a reaction. The law of conservation of energy states that the total energy cannot change in a reaction.

Cosmic rays. Energetic particles such as protons that come from outside the Earth's atmosphere.

Cosmology. The parts of astrophysics and astronomy having to do with the large-scale behavior of the universe and with its origin.

CP violation. An experimentally discovered, but not understood, phenomenon in the decay of neutral *K* mesons that violates some previously held ideas about the connection of particles to antiparticles and about time reversal.

Cryogenics. The science and technology of producing and using very low temperatures, even approaching absolute zero temperature.

DESY. Deutsches Electronen Synchrotron, the laboratory in Hamburg, Federal Republic of Germany, and its 6-GeV circular electron accelerator.

Deuterium. Heavy hydrogen, the nucleus of which contains one proton and one neutron.

DORIS. An electron-positron collider at DESY with a maximum energy of about 10 GeV.

Down. The distinguishing characteristic of one of the two lightest quarks, also called the *d* quark. The other light quark is the up or *u* quark.

Drift chamber. A particle detector in which the paths of charged particles produce tracks of ionized gas. Electrical signals from those tracks are detected and recorded, allowing the reconstruction of the particle paths.

Electromagnetic force or interaction. The long-range force and interaction associated with the electric and magnetic properties of particles. This force is intermediate in strength between the weak and strong force. The carrier of the electromagnetic force is the photon.

Electron. An elementary particle with a unit negative electrical charge and a mass 1/1840 that of the proton. Electrons surround an atom's positively charged nucleus and determine the atom's chemical properties. Electrons are members of the lepton family.

Electron volt. The amount of energy of motion acquired by an electron accelerated by an electric potential of one volt: MeV, million electron volts; GeV, billion electron volts; TeV, trillion electron volts.

Electroweak force or interaction. The force and interaction that represents the unification of the electromagnetic force and the weak force.

Elementary particle. A particle (piece of matter) that has no other kinds of particles inside of it and no subparts that can be identified. Hence the simplest kind of matter.

Elementary-particle physics. The area of basic science whose goal is to determine and understand the structure and forces of the most basic constituents of matter and energy.

Fermilab. Fermi National Accelerator Laboratory.

Flavor. A general name for the various kinds of quarks, such as up, down, and strange. Also sometimes applied to the various kinds of leptons.

Gamma rays. A term used for the energetic photons that are emitted in the decay of unstable particles and nuclei.

Gauge theory. A type of general theory of forces, modeled on the immensely successful modern theory of electromagnetism.

Generation. The classification of the leptons and quarks into families according to a mass progression. The first generation consists of the electron and its neutrino and of the up and down quarks. The second generation consists of the muon and its neutrino and of the charm and strange quarks. The third generation consists of the tau and its neutrino and of the bottom and expected top quarks.

GeV. (Giga electron volt) A unit of energy equal to one billion (10^9) electron volts.

Gluon. A massless particle that carries the strong force.

Grand unified theory (GUT). A hoped-for unification of the electroweak force with the strong force into a single gauge theory.

Gravitational force or interaction. The weakest of the four basic forces and the one responsible for the weight of matter and the motion of the stars and planets.

Graviton. A proposed massless particle that is assumed to carry the gravitational force.

Hadron. A subnuclear, but not elementary, particle composed of quarks. The hadron family of particles consists of baryons and mesons. These particles all have the capability of interacting with each other via the strong force.

HERA. An electron-proton circular collider being constructed at the DESY laboratory in the Federal Republic of Germany.

Higgs mechanism and particle. A mechanism that may explain the origin and value of the mass of all or some of the elementary particles. The mechanism includes a proposed set of particles called Higgs particles.

High-energy physics. Another name for elementary-particle physics. This name arises from the high energies required for experiments in this field.

IHEP. A 76-GeV circular proton accelerator in Serpukhov, USSR.

Intermediate vector boson. The general name for the W and Z particles that carry the weak force.

Invariance. A property of physical laws and equations such that they do not change when changes are made in reference or coordinate systems.

J. A particle made of a c quark (see *Charm*) and an anti-c-quark. It is also called the psi particle and is three times as massive as the proton.

Jet. A narrow stream of hadrons produced in a very-high-energy collision.

K meson or kaon. The next to the lightest meson. It is the lightest hadron that contains a strange quark.

KEK. A 12-GeV circular proton accelerator at Tsukuba, Japan.

LAMPF. An 800-MeV linear proton accelerator at Los Alamos National Laboratory, used for nuclear and elementary-particle physics.

LEP. A circular electron-positron collider with a maximum design energy of about 200 GeV being constructed at CERN, Switzerland.

Lepton. A member of the family of weakly interacting particles, which includes the electron, muon, tau, and their associated neutrinos and antiparticles. Leptons are not acted on by the strong force but are acted on by the electroweak and gravitational forces.

Lifetime. A measure of how long an unstable particle or nucleus exists on the average before it decays.

Linac. An abbreviation for linear accelerator.

Linear accelerator. In this type of accelerator, particles travel in a straight line and gain energy by passing once through a series of electric fields.

Luminosity.   A measure of the rate at which particles in a collider interact. The larger the luminosity the greater the rate of interaction.

Magnet.   A device that produces a magnetic field and thus causes charged particles to move in curved paths. Magnets are essential elements of all circular accelerators and colliders, as well as of many particle detectors.

Magnetic monopole.   A hypothetical particle that would carry a single north or south magnetic pole. All known particles with magnetic properties carry both a north and a south magnetic pole.

Mass.   The measure of the amount of matter in a particle and an intrinsic property of the particle.

Meson.   Any strongly interacting particle that is not a baryon. Mesons are composed of quark-antiquark combinations.

MeV.   (Mega electron volt) A unit of energy equal to one million electron volts.

Molecule.   A type of matter made up of two or more atoms.

Muon.   A particle in the lepton family with a mass 207 times that of the electron and having other properties similar to those of the electron. Muons may have positive or negative electric charge.

Neutrino.   An electrically neutral and massless particle in the lepton family. The only force experienced by neutrinos is the weak force. There are at least three distinct types of neutrinos, one associated with the electron, one with the muon, and one with the tau.

Neutron.   An uncharged baryon with mass slightly greater than that of the proton. The neutron is a strongly interacting particle and a constituent of all atomic nuclei except hydrogen. An isolated neutron decays through the weak interaction to a proton, electron, and antineutrino with a lifetime of about 1000 seconds.

Nucleon.   A neutron or a proton.

Nucleus.   The central core of an atom, made up of neutrons and protons held together by the strong force.

Particle.   A small piece of matter. An elementary particle is a particle so small that it cannot be further divided—it is a fundamental constituent of matter.

Particle detector.   A device used to detect particles that pass through it.

PEP.   An electron-positron circular collider with a maximum energy of 36 GeV, at SLAC.

PETRA.   An electron-positron circular collider with a maximum energy of 46 GeV, at DESY, Hamburg, Federal Republic of Germany.

Photon. A quantum of electromagnetic energy. A unique massless particle that carries the electromagnetic force.

Pion. The lightest meson.

Positron. The antiparticle of the electron.

Proton. A baryon with a single positive unit of electric charge and a mass approximately 1840 times that of the electron. It is the nucleus of the hydrogen atom and a constituent of all atomic nuclei.

PS. A circular proton accelerator with a maximum energy of 28 GeV at CERN, Switzerland.

Psi. A particle made of a *c* quark (see *Charm*) and an anti-*c*-quark and three times as heavy as the proton. It is also called the *J* particle.

Quantum chromodynamics (QCD). A theory that describes the strong force among quarks in a manner similar to the description of the electromagnetic force by quantum electrodynamics.

Quantum electrodynamics (QED). The theory that describes the electromagnetic interaction in the framework of quantum mechanics. The particle carrying the electromagnetic force is the photon.

Quantum mechanics. The mathematical framework for describing the behavior of photons, molecules, atoms, and subatomic particles. According to quantum mechanics, the forces between these particles act through the exchange of discrete units or bundles of energy called quanta.

Quarks. The family of elementary particles that make up the hadrons. The quarks are acted on by the strong, electroweak, and gravitational forces. Five are known, called up, down, strange, charm, and bottom. A sixth, called top, is expected to exist.

Relativistic. The term that describes particles moving with velocities close to the velocity of light.

Scattering. When two particles collide, they are said to scatter off each other during the collision.

Scintillation counter. A particle detector in which the passage of a charged particle produces a flash of light called scintillation light. That light, when detected, records the time at which the particle passed through the counter.

SLAC. Stanford Linear Accelerator Center in Stanford, California. Also refers to the electron linear accelerator there that is being rebuilt to have a total energy of 50 GeV.

SLC. Stanford Linear Collider, a linear electron-positron collider with an initial total energy of about 100 GeV being constructed at SLAC.

SPEAR. A circular electron-positron collider with a total energy of about 8 GeV at SLAC.

Sp̄pS. A circular proton-antiproton collider at CERN that uses the SPS accelerator there and has a total energy of about 600 GeV.

SPS. A circular proton accelerator with a maximum beam energy of about 400 GeV at CERN, Switzerland.

SSC. See *Superconducting Super Collider.*

Standard model. A collection of established experimental knowledge and theories in particle physics that summarizes our present picture of that field. It includes the three generations of quarks and leptons, the electroweak theory of the weak and electromagnetic forces, and the quantum chromodynamic theory of the strong force. It does not include answers to some basic questions such as how to unify the electroweak forces with the strong or gravitational forces.

Storage ring. An acceleratorlike machine composed of magnets arranged in a ring used to store circulating particles or to act as a collider. Sometimes a synonym for a collider.

Strangeness. The distinguishing characteristic of the third type of quark, also called the *s* quark. Each quark is characterized by a number of properties, including familiar ones like mass and electric charge and less familiar ones that were arbitrarily given names like charm and strangeness.

Strange particle. The name given to particles thought to contain just one *s* quark. The remaining quarks in strange particles are either *u* or *d* quarks.

Strong force or interaction. The short-range force and interaction between quarks that is carried by the gluon. The strong force also dominates the behavior of interacting mesons and baryons and accounts for the strong binding among nucleons.

Superconducting magnet. See *Superconductivity.*

Superconducting Super Collider (SSC). A design for a circular proton-proton collider with a total energy that could be as high as 40 TeV being developed in the United States.

Superconductivity. A property of some metals that when they are cooled to a temperature close to absolute zero, their electrical resistance becomes exactly zero. Magnets with superconducting coils can produce large magnetic fields while keeping size and power costs small.

Supersymmetry. A proposed theory of elementary particles in which a property of particles called spin is used. In most theories particles that differ in spin in some ways cannot be related. In this theory such particles can be related through a new proposed symmetry principle called supersymmetry.

Symmetry. A general property of many objects and physical systems whereby the object or system appears unchanged when looked at from different reference frames or coordinate systems. For example, a tennis ball has spherical symmetry because it always looks the same to us no matter how we move around it.

Synchrotron. A type of circular particle accelerator in which the frequency of acceleration is synchronized with the particle as it makes successive orbits.

Synchrotron radiation. Intense light or x rays emitted when electrons move in a circular orbit at relativistic speeds.

Target. The material, often liquid hydrogen, that is struck by the beam of high-energy particles in some types of elementary-particle physics experiments.

Tau. An elementary particle in the lepton family with a mass 3500 times that of the electron but with similar properties. There are positive and negative tau particles.

Technicolor. A proposed theory for explaining the masses of particles that postulates the existence of a new force.

TeV. (Tera electron volt) A unit of energy equal to one thousand billion (10^{12}) electron volts.

Tevatron. A complex of accelerator facilities and beam lines at Fermilab. The main facility is a circular proton accelerator with superconducting magnets (the first large accelerator to use such magnets) with a maximum energy of 1 TeV. An addition is being constructed so that this accelerator can be used as an antiproton-proton collider with a total energy of 2 TeV. On completion, this will be the highest-energy collider in the world.

Top. The distinguishing characteristic of the expected sixth type of quark, also called the truth quark or t quark. Each quark is characterized by a number of properties, including familiar ones like mass and electric charge and less familiar ones that were arbitrarily given names like bottom and top.

TPC. (Time projection chamber) A particle detector in which the position of the track of ionized gas left by a charged particle is detected by the time it takes for the electrons in the gas to move to the ends of the chamber.

TRISTAN. A circular electron-positron collider with a total energy of 60 to 70 GeV under construction at the KEK laboratory in Japan.

Unified theories. Theories of forces in which the behavior of different kinds of forces is described by a unified or single set of equations and has a common origin. For example, the electric and magnetic forces are unified in the theory of electromagnetism.

UNK. A complex of high-energy circular proton accelerators and colliders under construction at Serpukhov, USSR.

Up. The distinguishing property of one of the two lightest quarks, also called the *u* quark. The up and down quarks form the first quark generation.

Upsilon. A meson made up of a *b* quark and an anti-*b*-quark. It is approximately ten times as massive as the proton.

VEPP-4. A circular electron-positron collider with a total energy of up to 14 GeV at Novosibirsk, USSR.

W. The charged particle that carries the weak force, also called an intermediate-vector boson. Its mass is about 90 times the proton mass.

Weak force or interaction. The force and interaction that is much weaker than the strong force, but stronger than gravity. It causes the decay of many particles and nuclei. It is carried by the *W* and *Z* particles.

X rays. Photons produced when atoms in states of high energy decay to states of lower energy.

Z. The neutral particle that carries the weak force, also called an intermediate-vector boson. It is slightly heavier than the *W* particle, with a mass about 100 times the proton mass.

Index

223

European Center for Nuclear Research
(CERN), 175, 185, 186-187
defined, 214
proton-proton and proton-antiproton
colliders at, 113-114
European Committee on Future Accelerators (ECFA), 185
eV, *see* Electron volts
Evolution of universe, 3
Executive summary, 1-10
Extended families, 76-77

F

Federal government and elementary-par-
ticle physics, 182-185
Fermi National Accelerator Laboratory
(FNAL) (Fermilab), 174
facilities, 113
Tagged Photon Spectrometer, 147, 148
Feynman diagrams, 33, 35
Fixed-target accelerators, 41-42
electron, 205
experiments at, 40-41
proton, 88, 204
world's high-energy, 110
Fixed-target detectors, 133, 144-149
Fixed-target experiments, 5
detectors in, 133, 144-149
large or complex, 147
small or simple, 144-147
Flavor, defined, 216
Fly's Eye detector, 153
FNAL, *see* Fermi National Accelerator
Laboratory
Force unification, 26
Force-carrying particles, 27
Forces
four basic, 25-26
fundamental, unification of, 86
Funding for elementary-particle physics
research, 209, 212

G

Gamma rays, defined, 216
Gauge invariance, 85
Gauge symmetry, 85
Gauge theories, 3, 39, 72, 77
defined, 216
Generation, defined, 216

GeV (billion electron volts), 24
defined, 216
Global symmetry, 38
Glueball states, 88, 89
Gluon-gluon collisions, 91
Gluons, 3, 27, 30, 72
defined, 216
quarks interacting with, 72
Government, federal, and elementary-
particle physics, 182-185
Graduate education, 181-182
Graduate students, 189
Grand unified theories (GUT), 82, 159-
160
defined, 216
Gravitation, 13, 92
Gravitational force, 3, 25-26, 97
defined, 216
Gravitons, 14, 27
defined, 217
Gravity, quantum, 97
GUT, *see* Grand unified theories

H

Hadron-hadron storage ring colliders, 205
Hadrons, 2, 22, 64, 194
calorimeter, 139
defined, 217
family of elementary particles, 29-30
interactions of, 59, 87
jets of, 62-64
quark model of, *see* Quark model
secondary, 139
Heisenberg uncertainty principle, 24
HEPAP (High Energy Physics Advisory
Panel of the Department of Ener-
gy), 122, 183, 184, 185, 193
HERA, 111, 188, 208
defined, 217
HESYRL ring, 168
Higgs bosons, 91-92
light, 89, 91
Higgs field, 166
Higgs mass problem, 93
Higgs mechanism, defined, 217
Higgs particles, 67, 85-86, 121
defined, 217
High Energy Physics Advisory Panel of
the Department of Energy (HE-
PAP), 123, 183, 184, 185, 193

Triggers in detectors, 150
TRISTAN, 111, 116-117, 168, 208
  defined, 221
TRIUMF, 162
Two-beam accelerators, 130

U

UA1 detector, 133, 136, 137
Unification
  force, 26
  of fundamental forces, 86
Unification point, 79
Unified theories, 76-80
  defined, 221
  grand, *see* Grand unified theories
Universe
  evolution of, 3
  net baryon number of, 79
Universities Research Association
    (URA), 174
University-based research groups, 196-
    197
UNK, 111, 208
  defined, 222
Up quark, 29
  defined, 222
Upsilon particles, 30, 55
  defined, 222
  triplet states of, 57
URA (Universities Research Associa-
    tion), 174
UVSOR ring, 168

V

Vacuum fluctuations, ephemeral, 71-72
Vacuum polarization, 73
VEPP-2M ring, 168
VEPP-3 ring, 168
VEPP-4 ring, 112, 116, 168, 206
  defined, 222
Vertex detectors, 140-141
Volts, electron, *see* Electron volts

W

W particle, 3, 4, 5, 66, 222
Weak bosons, 66
Weak force, 3, 13, 22, 25-26, 106, 107
  defined, 222

X

X rays, 2
  defined, 222
Xi-star family of hadrons, 49

Z

Z particle, 3, 4, 5, 66, 222
Z^0 factories, 89-90
Z^0 particle, 67, 106
Zero electrical resistance, 103